THE BUG BOOK

"The right to be left alone—the most comprehensive of rights, and the right most valued by a free people."
—Justice Louis Brandeis, United States Supreme Court
Olmstead v. United States, 277 U.S. 438 (1928)

M.L. Shannon

THE BUG BOOK

Everything You Ever Wanted to Know About Electronic Eavesdropping . . . But Were Afraid to Ask

Paladin Press · Boulder, Colorado

Also by M.L.Shannon:

Don't Bug Me: The Latest High-Tech Spy Methods
Digital Privacy: A Guide to Computer Security
The Phone Book: The Latest High-Tech Techniques and Equipment for Preventing Electronic
    Eavesdropping, Recording Phone Calls, Ending Harassing Calls, and Stopping Toll Fraud

The Bug Book: Everything You Ever Wanted to Know
about Electronic Eavesdropping . . . But Were Afraid to Ask
by M.L. Shannon

Contents

CONTENTS

Warning

The Bug Book was written *for information purposes only*. Neither the author, publisher, distributor, seller, or any person or organization involved in promoting this book assumes, or will assume, any responsibility for the use or misuse of the information contained herein; nor will they be responsible for the consequences thereof, either civil or criminal. If you have doubts or questions as to what is or is not legal, please consult an attorney.

The computer system used to write, produce, and typeset this book is linked to, and is part of, the electronic mail system, including but not limited to the Internet. All of the research material on this computer system, and in this book, including all future editions and revisions, is being prepared, or was prepared, for public dissemination and is therefore work product material protected under the First Amendment Privacy Protection Act of 1980 (USC 42, Section 2000aa). Violation of this statute by law enforcement agents may result in a civil suit as provided under Section 2000aa-6. Agents in some states may not be protected from personal civil liability if they violate this statute.

The possession and/or use of some of the devices described in this book may be unlawful. Monitoring cellular radio and cordless telephone frequencies and some other frequencies and types of transmissions is a violation of federal law and local laws in some areas. Again, if you have any doubts as to what is or is not legal, you are advised to consult an attorney.

Also be aware that if you bug the wrong people, jail may be the least of your worries.

Preface

The Bug Book is about bugs: hidden transmitters that can broadcast your personal and business conversations to industrial-strength spies, government agents, or perhaps the next door neighbors—people who can listen to and record your most intimate and important words, and perhaps use this information to their own advantage. For blackmail, perhaps. Or to get an advantage over you in business matters. Or to attempt to destroy you: wreck your marriage, cost you your job, or whatever else.

The Bug Book is about how these devices work, how effective they are, and how to find them and deal with them.

The Bug Book is also about how to use this technology in one's own defense, should it become necessary.

The Bug Book is not intended for corporate security personnel or private investigators, although they will probably find it useful. It was not written for federal agents, even though the "Alphabet Soup" agencies will purchase it. No, *The Bug Book* was written specifically for the general public—the machinist in Des Moines, the secretary in Omaha, the plumber in Sacramento, the cab driver in Phoenix, and the houseperson in Albany. My intent is to present this information, including the technical stuff, in a way that the "average person" can understand and use.

I either know or believe the technical information in this book to be true and accurate, based on two years of college in electronics, employment as a technician for a company that made surveillance and countermeasures gear as well as for a professional countermeasures company, a lifelong interest in communications, and some personal experiences with surveillance. All of the case histories are, unless otherwise stated, based on real-life events. Some of the details, such as names and locations, have been changed for obvious reasons. Some of the details have not been changed for my own reasons. You know who you are. At least some of you do . . .

BUT, IS THIS BOOK NECESSARY?

In the Spring of 1991 I completed my first book, which after some thought I named *Don't Bug Me*. I didn't have a publisher, and I had no idea whether or not I would be able to sell it. But it was done; I'd written a book. It was photocopied and spiral bound at a local copy shop in San Francisco. The typesetting wasn't the greatest, and the graphics, from a dinosaur Kodak machine, left a lot to be desired. Initially, I sold it mainly through the now defunct Spy Factory chain. Hoping for the best but expecting the worst, I sent off copies to several publishers, one of which was Paladin Press. A while later, I heard from them: they wanted it. The ultimate rush! A celebration of this momentous occasion was in order. So, I set off to quaff a pint or two at Edinburgh Castle, a Scottish bar that features Very British fish 'n chips cooked in a Korean kitchen. Only in San Francisco. While sucking up suds, naturally I was telling anyone who would listen about my great fortune. Although there were many well wishers and a few free Heinekens, not everyone was impressed.

One old guy got all bent out of shape and started bellering at me: "Why, you're publishing information that criminals and drug dealers can use to evade the law. You have no right . . ."

That was the wrong thing to say. I told him that it was in no way my intention to do anything to help "criminals and drug dealers evade the law." "While I am well aware that some of the information *could* be used by such people for that purpose," I said," I believe this would be on a very small scale."

"I believe that people should have the right to know," I told him, adding that the government—the federal law enforcement and intelligence agencies—do not agree. On and on I went about how they want to take away that right, monopolize spying, and exercise total control over all forms of communication—and how in the last few years they have made great advances in their war on knowledge and privacy.

Soon a number of people in the pub were listening intently. I ranted about how the government is sticking their electronic noses into people's private business without a warrant in deliberate violation of existing laws, because they know they will always get away with it.

Decency, security and liberty alike demand that government officials shall be subjected to the same rules of conduct that are commands to the citizen. In a government of laws, existence of the government will be imperiled if it fails to observe the law scrupulously. Our Government is the potent, the omnipresent teacher. For good or for ill, it teaches the whole people by its example. Crime is contagious. If the Government becomes a lawbreaker, it breeds contempt for law; it invites every man to become a law unto himself; it invites anarchy. To declare that, in the administration of the criminal law, the end justifies the means—to declare that the Government may commit crimes in order to secure the conviction of a private criminal—would bring terrible retribution. Against that pernicious doctrine this Court should resolutely set its face.

—Justice Louis Brandeis
Olmstead v. United States

I raved about big business, corporate America, which will always be able to get surveillance equipment, as will some "PI" types, some of whom will use it unlawfully. "Why should we be forbidden the right to defend ourselves against unlawful surveillance, and the right to use it to fight back when there is no other way?" I asked. "Privacy is dying at the hands of government," I told them. I asked the people who were listening what they would do if they thought someone had them under surveillance; how would they defend themselves? How would they even know about it? My question resulted in silence, everyone staring at me with blank expressions. Like most people, they had never even thought about it.

"This," I told them, "is what *Don't Bug Me* is all about." That is why I wrote it.

Damn right this book is necessary. It was necessary then, and it is even more necessary today, eight years later, as this sequel, *The Bug Book*, is published. And as far as the "right," it is more than that. It is a responsibility.

Introduction
Who's (Bugging) Who?

Who's (bugging) who? Anyone who has the need (or just the desire), who possesses the equipment, and who has access to the target area. Motive, method, and opportunity.

Most bugging is industrial: big corporations (and some not-so-big businesses) spying on their competitors.

In a lawsuit between two giant corporations, millions of dollars may be at stake. At some point in discovery, each side shares their list of witnesses. But if one of the law firms were able to get advance notice of these witnesses or other parts of the game plan, it would give them an advantage over the opposition. So they spy.

The chemical formula for a new Super-Healthy Vita-Charged (46-percent added sugar) breakfast cereal or the design details of a new car or computer or the marketing plan for an over-the-counter cold remedy could be worth big bucks to a competing company. Nothing personal, of course; it's just business. So they bug each other's offices and boardrooms and bathrooms. But why stop there? An employee who is privy to this classified information might be bribed or blackmailed into selling company secrets. So, they bug bedrooms. Much easier than boardrooms and probably more entertaining to whoever is listening. Using technology that has been around for some years, they may be able to tap phones by remote control. Connect a personal computer to the telephone company DMS electronic switching system, and dial in the number you want to tap. A slight oversimplification, but basically that's how it works. Get the goods on some poor guy in marketing, apply a little pressure. . . . This is the world of industrial espionage, where profit isn't everything; it's the only thing. A few thousand bucks spent on surveillance is a good "investment." So they spy.

Then we have good old domestic spying—wives and husbands, boyfriends and girlfriends checking up on each other. Once limited to listening in on the old telephone party lines or pressing an ear to the walls, it has evolved to a science. Neighbors spy on neighbors using directional microphones or tuning in on their cellular and cordless phones. They hide bugs under the bed and connect tape recorders to the phones. And sometimes the result is not what was expected: "Hey, I'm on the radio!"

Linda came home from work one afternoon, and brought with her something that was intended to amuse her husband and some friends who had been invited over for a dinner party.

After the lamb chops and a mediocre soufflé and the kids were put to bed, everyone gathered in the living room. But this night was not to be the usual boring charades and strip poker. Linda produced her surprise, and after tuning the stereo to an unused frequency, started playing disk jockey. This was quite a few years ago, when such things were not so well known. What she had was a wireless microphone. Everyone got a kick out of this new form of entertainment, others started buying the homemade devices, and it wasn't long before one of the big mail order companies came out with Mr. Microphone and a thousand commercials to tell the world about it.

Later that night, the party ended. The guests left, and Linda and her husband went to bed. The transmitter was on a coffee table, where the last person to play DJ had left it. Linda didn't notice that it was missing the next morning as she left for work, and by the time she got home the transmitter was forgotten.

Sometimes kids do innocent things they think are clever, not intending to harm anyone. In this case, they were too young to really understand that hiding the transmitter in their parents' bedroom was not a nice thing to do—especially since that night, Linda and her husband got into a heated argument. At first the kids thought it was funny, but then Linda's husband said something that he could never take back. Something that the kids heard. Something that planted the seeds of doubt and finally resulted in a divorce.

Not the most entertaining of stories, but it makes a point—something that anyone who is considering using surveillance against another should think about, even if it is in self-defense. The consequences can be devastating. The information obtained can break up families, cause people to lose their jobs, and so many other things, just because someone bugged them. Think about it.

Then there is the federal government. I read somewhere that there are 17,000 separate agencies, most of which the public has never heard of, and which probably have little interest in spying. Other agencies are very interested; electronic spying is part of their everyday jobs. A big part. The National Reconnaissance Office (NRO) with its high-resolution spy satellites, the NSA (No Such Agency), where spying is the only thing it does, the Central Intelligence Agency (CIA), which spies whenever it suits it (which is most of the time), and the Federal Bureau of Investigation (FBI), where the J. Edgar Hoover mentality is alive and well. They are obsessed with tapping every phone in the country, including granny Yokum. Have you ever called the embassy of another country, for whatever reason? The fibbies know about it. Ever make an overseas call? The NSA knows about it and may have recorded it. Ever call the White House to tell them how much you (love, hate, whatever) the present administration? The Secret Service has a file on you.

And now the Internet, for many years a well kept secret, has become accessible to the general public, and suddenly tens of millions of people have electronic mail. And the feds want desperately to read this E-mail. Your E-mail. And your granny's E-mail. And at one time or another, they probably have. This is, of course, necessary. After all, maybe your granny has sacks of ammonium nitrate fertilizer in her basement, which means she is probably a terrorist.

This is an enormous task, reading these millions of messages, so the feds have specially designed processor in memory supercomputers (PMS). Special software scans the text, looking for certain words and phrases, and if any of these are found, the sender (and probably the recipient) gets his or her name in a file. Hope Granny didn't tell anyone she plans to fertilize her begonias, or if she did, that she was careful in her choice of words.

In the years since *Don't Bug Me* was published, technology has changed. New and better equipment has been developed, some of which defeats all but the most experienced and determined countermeasures technician. The feds have development laboratories, perhaps at Langley or Ft. Meade, and such places (no guided tours here) where they design and make the best equipment that exists. Much of the stuff they use to spy won't be found in mail order catalogs, and the salesperson at your local black market spy shop won't be much help if you ask him or her about it. And if you are the average businessperson, homemaker, or working stiff, you will not likely encounter this super-sophisticated government spy stuff.

But in spite of the new technology, some things do not change. Techniques, the basic principles of electronic surveillance, remain essentially the same. For example, if you can not get a microphone close enough to a person to pick up the sound of his or her voice, then a surveillance transmitter is useless. Makes no difference if it is a $12 wireless microphone or a $1,000 spread spectrum transmitter. If you can't get access to a phone line, you can't physically tap it. So, no matter who might be trying to bug you; spouse, boss, competitor, or Big Brother, and no matter how sophisticated the equipment, there are certain principles—laws of physics and electronics—that never change. There are also elements of human nature that do not change. Ergo, the principles of using surveillance equipment, and of defeating it, are much the same. In this book you will learn a great deal about them.

WHAT'S NEW?

The Bug Book was first published by Lysias Press in 1994 but has been out of print for several years. Since then, I have completely rewritten it and have added much new information. The culmination of that effort, this book, is close to twice the size of the first edition and has dozens of photographs, where the original edition didn't have any. Some of the new material includes:

Computer-aided scanning (CAS). Some basic info on CAS—how to use a computer to control a scanner or communications receiver—is in *The Phone Book*. Here, I have elaborated and included a review of Radio Max. This program, along with a good scanner, can be used as an inexpensive way to search for bugs.

Phone-phreaking terms and tricks. A closer look inside the telco, some exotic terms explained, things you can do and things you don't want to do. Plus, a wiretap detecting device you can build, with no electronics experience, for less than $20. Finally, an interesting story: something that happened to me recently.

Microwave surveillance. Surveillance transmitters that operate in the microwave portion of the spectrum may not be detected in a countermeasures sweep because the equipment used may not cover this high area. While not a project for the beginner, building a microwave transmitter that operates in the 10 gigahertz (GHz) band and has a potential range of several miles is possible. And you can use a modified radar detector to receive the signal. There are also commercial microwave transmitters that can be used for surveillance, but they are quite expensive.

Spread spectrum. This is the "new technology" for many forms of communications, and surveillance is no exception. It has many advantages over traditional FM: the range is greater, the signal is less affected by obstructions, and it is difficult, not impossible, to detect with most countermeasures equipment.

Infrared transmitters. They are not as easy to use as radio frequency (RF) bugs, but they are also not as easy to detect. The range is limited, but they do indeed work. They're not cheap, but they are available to anyone who wants them from retail stores across the country. Very little assembly required.

A surveillance FAQ: frequently asked questions. This was taken from my Web site, fusionsites.com, and includes information on surveillance methods and equipment other than RF "bug" transmitters.

Sources: where to get things. What with the feds having closed down most of the "spy" shops and mail order dealers, surveillance devices aren't as readily available as they were a few years ago. So this chapter from *Don't Bug Me* had to be rewritten completely. Listed here are sources of some bugs, as well as some other products with which one can improvise.

An index. The reviews of my books have been very good, but one complaint kept popping up: There is no index. ("An otherwise excellent book, but we hope the author will consider including an index in his next work.") I have, and Paladin Press agrees that this is a good idea.

HOW MUCH SPYING IS GOING ON?

According to "Electronic Eavesdropping & Industrial Espionage," a newsletter published by Kevin Murray of the Clinton, New Jersey-based professional countersurveillance company Murray Associates,

> Due to the covert nature of spying, we will never know for certain [how much of it is going on]. Fortunately, however, we can use the failed espionage attempts as a gauge. They reveal over and over again that the problem does exist. Also, the plethora of electronic surveillance equipment being openly sold in spy shop stores and executive toy catalogs gives us a good indication of the magnitude of electronic eavesdropping. Word filtering back through the press and from electronic eavesdropping detection specialists can make you a believer. It's happening on a daily basis.

And while the equipment is no longer "openly sold in spy shop stores" it can still be had.

Another thing to consider is the manufacturing of "restricted" surveillance devices. While this doesn't give any actual numbers, it is another indicator that there is a lot of spying going on. One of the world's largest manufacturer of electronic surveillance devices is Westinghouse/Audio Intelligence Devices (WAID). At least that's what it claims in its catalog. The company has a 75,000-square-foot facility in Coral Springs, Florida, that includes R&D and manufacturing departments, a complete machine shop, a surveillance van, assembly area, classrooms, a cafeteria, and hundreds of full-time employees. Its catalog, which is several hundred pages, lists a fascinating collection of goodies that would please even the most discriminating of spies. WAID doesn't deal with the general public. It won't send you its catalog unless you are a law enforcement officer or government spy, and the request has to be on official letterhead. And WAID products aren't cheap. A few examples:

- A crystal-controlled VHF 50-megawatt (MW) transmitter built into the magazine of a "standard" 9mm pistol. There is still space for three rounds, which can be fired without affecting the transmitter. However, I suspect it might affect the ears of whoever is listening. Only three grand.
- A 250-MW flat-pack transmitter, 5 x 3 1/8 x 3/8 inches with 6- to 10-hour battery life for $1,095.
- A 120-MW baseball cap transmitter with battery life of 3 hours for $1,320.

WAID also makes a large assortment of transmitters disguised as Walkman-type radios, pagers, and working electrical outlets; night vision equipment, audio equalizers, Bird Dog® vehicle tracking systems, and complete surveillance vans. Given the size of WAID's facility, it must make quite a few of these goodies.

PK Elektronik in Hamburg, Germany, is another large manufacturer of surveillance devices. It has a staff of more than 1,000 and occupies a 14-story office building, which is pictured in its catalog. Its beautiful 268-page catalog is printed on expensive glossy paper (it weighs about four pounds) and is full of very nicely done photography—hundreds of pictures of PK's vast array of products.

Unfortunately, it's not easy to get your hands on one of these catalogs. PK denied my initial request, stating in a letter that the company's products and related information "are restricted to government and law enforcement agencies and not available to the general public." However, a government or law enforcement type who orders this catalog will find that it contains transmitters that work on many frequencies, from VHF to microwave, that are built into electrical outlets, ash trays, picture frames, nonworking light bulbs, flower vases, and calculators, to name a few. There are even solar-powered transmitters. PK, like WAID, is not cheap; but, also like WAID, it makes very high-quality products. (So I am told.) Add to this such manufacturers as Cony in Japan, Lorraine in England, and others in Maryland and Virginia that I have never heard of, and there are a lot of people making a lot of spy stuff.

And someone is buying it. Someone? People are buying it. According to Murray Associates,

> One company used to sell miniature wireless microphones via mail order. Their ads appeared in the back of electronic magazines. To cover their advertising and product costs, they must sell approximately 50 transmitters per month, per ad. Any additional units sold represent profit. It must

be very lucrative. The ad had run every month for 18 years, and appeared in at least three magazines. Just to cover costs . . . 32,400 bugs were sold.

. . . And they probably aren't being used for conversation pieces. (Pun not intended.)

The federal government is buying it. City, state, and county police agencies are buying it. People are buying it.

Motive, method, opportunity.

The police and the attorneys general often complain that bleeding-heart-liberal judges refuse their requests. In the 11 years from 1982 to 1993, only seven applications for a surveillance warrant were denied. That's less than one-tenth of one percent. From 1993 to 1997 the percentage was zero. Every request was granted.

STATISTICS

As Murray states above, one can get a general idea of how much spying is going on in the private sector, but in the case of the government, more specific information is available. The reporting requirements of the Omnibus Crime Control and Safe Streets Act dictate that each federal and state judge is required to file a written report with the Director of the Administrative Office (AO) of the United States courts on each application for an order authorizing the interception of a wire, oral, or electronic communication (Title 18 U.S. Code 2519[1]). . . . Prosecuting officials who applied for interception orders are required to submit reports to the AO each January on all orders that were terminated during the previous calendar year.

These reports, naturally, aren't going to state who was being spied upon, but they do report the numbers, which are published, every year, by the Office of the U.S. Courts in a free book called simply *The Wiretap Report*. The statistics that follow are from the 1998 edition of that book.

The evil incident to invasion of the privacy of the telephone is far greater than that involved in tampering with the mails. As a means of espionage, writs of assistance and general warrants are but puny instruments of tyranny and oppression when compared with wiretapping.

—Justice Louis Brandeis
Olmstead v. United States

In 1998, the total number of applications was 1,329, up from 1,094 in 1997. Of these applications, all but two were approved. In the last 11 years, less than one-tenth of one percent were denied. The feds know which judges to go to . . .

- Of these applications, 566 were to federal judges and 763 to state magistrates. In 1987, federal surveillance installations were only 35 percent of the total. In 1998 it was 42.6 percent.
- In 1998, the highest number of orders by state were New York (304), Pennsylvania (105), New Jersey (70), and Florida (57). These four states accounted for 70 percent of all state installations. Incidentally, only 42 states have wiretap laws, the eight without being Alabama, Arkansas, Kentucky, Maine, Michigan, Montana, South Carolina, and Vermont.
- Authorizations are granted for a period of 30 days but can be renewed as many times as the agency desires. In 1997, 1,028 extensions were granted. The average was 28 days; the longest federal case was a pager intercept in Arizona that lasted 430 days, and the longest was in New York, lasting more than 5 years.

Table 2
Intercept Orders Issued by Judges During Calendar Year 1997

Reporting Jurisdiction	Number of Intercept Orders					Number of Extensions	Avg. Length (in Days)		Total Number of Days in Operation	Place/Facility Authorized in Original Application						
	Authorized	Amended	No Prosecutor's Report	Never Installed*	Installed*		Original Authorization	Extensions		Single-Familiy Dwelling	Apartment	Multi-Dwelling	Business	Roving	Combination**	Other
TOTAL	1,186	10	73	19	1,094	1,028	28	28	48,871	273	108	1	78	12	185	529
FEDERAL	569	3	-	6	563	560	30	29	29,055	118	44	-	30	4	113	260
ARIZONA																
MARICOPA	6	-	-	-	6	4	30	30	232	1	-	-	-	-	3	2
CALIFORNIA																
AMADOR	1	-	-	-	1	-	1	-	1	-	-	-	-	-	-	1
FRESNO	1	-	-	-	1	-	30	-	29	-	-	-	-	-	-	1
LOS ANGELES	24	-	-	-	24	13	30	30	911	6	-	-	-	-	5	13
MONTEREY	1	-	-	-	1	1	18	30	47	-	-	-	-	-	-	1
SONOMA	1	-	-	-	1	-	30	-	30	1	-	-	-	-	-	-
COLORADO																
EAGLE	1	-	-	-	1	-	18	-	13	-	-	-	-	-	-	1
JEFFERSON	1	-	1	-	-	-	2	-	-	-	-	-	-	-	-	1
MESA	2	-	-	-	2	-	30	-	48	2	-	-	-	-	-	-
CONNECTICUT																
HARTFORD	4	-	-	-	4	1	15	15	51	2	2	-	-	-	-	-
LITCHFIELD	4	-	1	-	3	-	15	-	27	4	-	-	-	-	-	-
FLORIDA																
1ST JUDICIAL CIRCUIT (ESCAMBRIA)	5	-	-	-	5	-	30	-	82	3	-	-	-	-	1	1
2ND JUDICIAL CIRCUIT (LEON)	15	-	-	-	15	5	25	30	499	8	-	-	3	-	2	2
4TH JUDICIAL CIRCUIT (DUVAL)	5	-	-	1	4	-	30	-	71	3	-	-	-	-	-	2
5TH JUDICIAL CIRCUIT (LAKE/MARION)	1	-	-	-	1	-	30	-	29	-	-	-	-	-	1	-
11TH JUDICIAL CIRCUIT (DADE)	14	-	-	-	14	4	30	30	500	5	-	-	1	-	3	5
13TH JUDICIAL CIRCUIT (HILLSBOROUGH)	4	-	-	-	4	1	30	11	128	-	-	-	-	-	3	1
15TH JUDICIAL CIRCUIT (PALM BEACH)	2	-	-	-	2	-	30	-	46	2	-	-	-	-	-	-
17TH JUDICIAL CIRCUIT (BROWARD)	2	-	1	-	1	-	30	-	29	-	-	-	-	-	2	-
18TH JUDICIAL CIRCUIT (BREVARD/SEMINOLE)	5	-	-	-	5	5	30	30	157	2	-	-	3	-	-	-
19TH JUDICIAL CIRCUIT (SAINT LUCIE)	4	-	2	-	2	1	30	30	49	4	-	-	-	-	-	-
GEORGIA																
BIBB	5	-	-	-	5	1	20	10	80	4	1	-	-	-	-	-
CHATHAM	1	-	-	-	1	-	20	-	10	1	-	-	-	-	-	-
FLOYD	2	-	-	-	2	2	20	20	60	2	-	-	-	-	-	-
GWINNETT	3	-	-	-	3	1	20	20	60	1	1	-	-	-	1	-
ROCKDALE	6	-	-	-	6	1	20	20	114	1	-	-	-	-	1	4
TROUP	1	-	-	-	1	1	20	21	14	-	-	-	-	-	1	-
ILLINOIS																
KENDALL	1	-	-	-	1	1	10	10	10	1	-	-	-	-	-	-

These two pages are samples from *The Wiretap Report*, published by the Office of the U.S. Courts. On this page are listed the total number of surveillance orders issued for the year to the feds, as well as some of the states. Page 7 shows the type of surveillance used, again for federal agencies and some of the states.

THE BUG BOOK

Table 6
Types of Surveillance Used, Arrests, and Convictions for Intercepts Installed
January 1 Through December 31, 1997*

Reporting Jurisdiction	Orders for Which Intercepts Installed	Phone Wire	Microphone Eavesdrop	Electronic	Combination**	Number of Persons	
						Arrested	Convicted
TOTAL	1,094	756	35	206	97	3,086	542
FEDERAL	563	415	16	75	57	1,765	294
ARIZONA							
MARICOPA	6	3	-	1	2	87	6
CALIFORNIA							
AMADOR	1	1	-	-	-	3	-
FRESNO	1	-	-	1	-	-	-
LOS ANGELES	24	8	-	15	1	28	-
MONTEREY	1	1	-	-	-	3	2
SONOMA	1	1	-	-	-	1	1
COLORADO							
EAGLE	1	1	-	-	-	1	1
JEFFERSON	NR	-	-	-	-	-	-
MESA	2	2	-	-	-	14	10
CONNECTICUT							
HARTFORD	4	4	-	-	-	-	-
LITCHFIELD	3	3	-	-	-	7	3
FLORIDA							
1ST JUDICIAL CIRCUIT (ESCAMBRIA)	5	3	-	2	-	-	-
2ND JUDICIAL CIRCUIT (LEON)	15	13	1	1	-	20	-
4TH JUDICIAL CIRCUIT (DUVAL)	4	3	-	1	-	5	-
5TH JUDICIAL CIRCUIT (LAKE/MARION)	1	1	-	-	-	17	2
11TH JUDICIAL CIRCUIT (DADE)	14	12	-	-	2	56	-
13TH JUDICIAL CIRCUIT (HILLSBOROUGH)	4	-	-	1	3	23	2
15TH JUDICIAL CIRCUIT (PALM BEACH)	2	2	-	-	-	2	-
17TH JUDICIAL CIRCUIT (BROWARD)	1	-	-	-	1	-	-
18TH JUDICIAL CIRCUIT (BREVARD/SEMINOLE)	5	5	-	-	-	12	-
19TH JUDICIAL CIRCUIT (SAINT LUCIE)	2	2	-	-	-	14	-
GEORGIA							
BIBB	5	5	-	-	-	11	4
CHATHAM	1	1	-	-	-	-	-
FLOYD	2	2	-	-	-	-	-
GWINNETT	3	2	-	-	1	17	-
ROCKDALE	6	6	-	-	-	11	-
TROUP	1	1	-	-	-	7	-

- The average federal cost was $73,404 up from $72,390 in 1997; the average state cost was $37,137.
- The breakdown of the location of surveillance installations was as follows: private homes, 23 percent; apartments, 9 percent; and businesses, 7 percent. The rest were "roving," which target an individual person, no matter where he or she is located, rather than a particular location; and "other," meaning cell phones, pagers, and RF transmitters.
- The type of crime for which surveillance was used for in 1998 is broken down as follows:

Narcotics	955
Racketeering	153
Kidnapping	5
Homicide	53
Larceny	19
Bribery	9
Gambling	93
Extortion	12
Other	30

The FBI claims that the new digital phone systems prevent it from being able to tap phones legally. A Freedom of Information Act request by the Electronic Privacy Information Center (EPIC) reveals that there is not a single instance where this is true (details at http://www.epic.org/).

Nevertheless, the fibbies got what they wanted with the new laws passed in August 1999.

J. Edgar Hoover was quoted in a newspaper in the thirties as saying, "If a child were kidnapped, wouldn't you want [law enforcement] to know the location where a [ransom] call was made?" Interesting, since surveillance was used in only five such cases.

Now, here is something interesting to ponder:

- WAID makes dozens of high quality, expensive transmitters.
- It costs a lot of money to design and produce these devices.
- WAID sells only to law enforcement agencies
- Law enforcement uses (legally) only a handful of transmitters per year.

How, then, can WAID afford to continue producing these exotic devices? Maybe law enforcement agencies are stocking up on bugs. Or, maybe they are using them unlawfully.

QUIET VICTIMS

We know that Surveillance Happens. We know that much of it is unlawful. But think about it: when was the last time you heard, through the media, about anyone being arrested for electronic spying? True, there are the semiannual "CIA Spook Sells Secrets to Soviets" stories, and the occasional "Citizen Catches Celebrity Cellular Conversation," but as far as the day-to-day "bugging thy neighbor," industrial espionage, and the like, you rarely hear a peep. There are two reasons for this, either or both of which will usually apply. First of all, the spies rarely get caught. (Maybe they have read my books.) And second, even if they do, the victim is unlikely to get law enforcement involved because of the bad publicity. I can see it now on Yahoo News: Wexler's Widget Works in Wortley, Wisconsin, announced today that a former disgruntled (they always call them disgruntled in the media) employee had placed bugs in the Wexler's boardroom and had been recording their minutes for months. And just below that: Wexler's Widget Works (NASDAQ WWW) plunges to a record low of $12 a share from a 52-week high of $83.42.

A big law firm involved in a huge civil action could be set back years and incur costs of millions, not to mention the loss of clients, if word were to get out (so to speak) that it had been successfully bugged. No one wants to talk about it. Consequently, no one hears about it. There is one other thing, should you be thinking about a career as a wireman (wireperson)—something else you don't read about in the papers or otherwise hear about: The reason so many bodies are found floating in the Hudson River in the spring. Bug the wrong people and get caught, and you just might end up there.

SURVEILLANCE AND THE LAW

How is anyone supposed to understand surveillance law? How are people to know what they can legally do and what they cannot? Laws and more laws are being passed so often that even lawyers have trouble keeping up with them. Many of these laws are so vague that just about anyone could be arrested by federal agents for possession of anything that could be used as a surveillance transmitter. And, as we enter the era of so-called cyber-terrorists, it won't be long before a person could be arrested just for possessing a computer. An ordinary computer. ("Well, no, your Honor, the defendant didn't actually commit a cyber-crime, but he owned a computer and he 'might have' so we ask that he be held without bail.") It has come to the point where a person can be taken into custody and held indefinitely without bail just because a federal agent is willing to state, in court, that this person "might" violate a law.

There are devices that are advertised as "bugs" and "surveillance transmitters" and kits that, when assembled, amount to the same thing. Is this where the government draws the line—anything advertised as "spy" instruments? Anything in kit form that could be used as such? What about the many other things that can be used for surveillance, even though their design is not such that they are primarily useful for spying on people? Title 18, U.S. Code, specifies that anything used for "the surreptitious interception of any wire, oral, or electronic communication" qualifies as a bug.

Every person who owns a "baby monitor" or two-way radio would be potentially violating the Omnibus Crime Control and Safe Streets Act (part of Title 18). Wireless microphones, as used by entertainers on stage and referees in football games, are apparently legal. I haven't heard of the feds raiding Rolling Stones concerts, no matter how much people like Dan Quayle dislike such "lack of family values."

As far as how (if and when) these laws are enforced, I suspect they are much like any other laws. What often matters is intent or probable intent. If you are a construction worker on the job and are using a bolt cutter to cut steel reinforcement bars (rebars) to be used in the construction of a building foundation, then the police aren't likely to take you to jail if they see you using it. On the other hand, if the cops see you hanging around in the alley behind a jewelry store at 3 A.M., and you have said bolt cutter with you, suddenly it becomes a burglary tool. This is intent, not the actual use of what may or may not be an unlawful surveillance device.

If you are using a transmitter (or any other electronic device) to listen to anyone without his or her knowledge and consent, then there is little doubt that this is a violation of federal, and probably local, laws. I think the key here is the "reasonable expectation of privacy," but what constitutes "reasonable" I do not know. Certainly, this applies to people in their own homes, but not entirely to their place of employment.

But what if you have an intercom set up at your front door, and it is always turned on, always listening? If it is used as a security device to protect your home or property to which you legally have access, then it is probably legal, providing it will not overhear anyone who also has legal access to the area. If you set it up in your office after closing, and someone breaks in and you hear them, this is probably legal. Using it as a wireless baby monitor is also, I believe, legal. Probably depends on who the "baby" is. Again, ask a lawyer.

FCC Law

The Federal Communications Commission (FCC) rules allow the use of low-power transmitters without a license on certain frequencies, which are listed in Title 47 of the Code of Federal Regulations (CFR) available at most any public library and on the Internet. This is all written in the federal government's usual legalese and is not easily comprehensible to ordinary mortals, but it includes those frequencies used for cordless phones, auditory aids for the hearing impaired, and wireless modems. (A list is included in Appendix D.)

Title 47, Part 15, subparts 118 and 119 concern the amount of power that can be used. As near as I can understand it, for FM wireless microphones, the combination of battery voltage and antenna length is to be such that the range is no more than 75 feet when received on a standard FM radio. Most commercial systems advertise a range far exceeding 75 feet, and if you are listening with a communications receiver, which is much more sensitive, you would be able to hear it at a much greater distance.

For cordless phones and other low-power devices, the radio frequency output can not be more than 100 MW, and the antenna has to be attached to the device (in other words, not mounted on top of a 200-foot tower) and not more than one meter (m) in length. Good for a block or so, depending.

Cordless phones, baby monitors, and wireless microphones are required to be type accepted by the FCC. If a device is type accepted, the following will be printed on the outside: "This device complies with part 15 of FCC rules. Operation of this device is subject to the following two conditions: [1] This device may not cause harmful interference. [2] This device must accept any interference that may cause undesired operation." With this will be a number, such as FCC ID CCT9P6157T. But getting type acceptance takes some doing.

From the FCC Equipment Authorization Program for RF devices, Office of Engineering Technology, May 1987:

> Part 2 sect 2.801 FCC rules says that it is unlawful to import non type accepted electronic devices. The reason for this is to lessen RFI (Radio Frequency Interference). To get acceptance of a low-power transmitter that operates on the commercial FM band, obtain and fill out FCC form 731 and send it to the FCC at 1919 M St. NW, Washington, DC 20554. One sample transmitter has to be included with the application. The fee is $1300.00.

To check on the type acceptance of a device, the government has the Public Access Link (PAL) BBS, which is open to the public (call 301-725-1072).

Back to the Future. In the first edition of *The Bug Book*, I wrote the following:

> Someone who claimed to be a former federal agent once said to me during an interview: "These people think they're being smart, selling the things in kit form. We could bust the lot of them anyway, and even if we didn't get a conviction, the cost of a trial in fed[eral] court would break them."
>
> I asked him, "You mean that even though they are not breaking any law, and you know that they aren't breaking any law, you might arrest them anyway?"
>
> "Look, these people are selling bugs, and they ought to be shut down . . ."
>
> Such is the awesome power of the government over the people, and one of these days it is going to happen. Get 'em while you can.
>
> That's what I wrote several years ago, and as we now know, this is just what has happened. Massive raids on The Spy Factory put the corporation and the 14 retail stores out of business and the owners in the slammer. Deco and Xandi no longer sell transmitters, and Sheffield is thinking about relocating to Ireland. (But WAID is alive and well.)

WHAT, ME? WORRY?

With apologies to Alfred E. Neuman, ignorance is not by any means bliss. We have seen, so far, that Surveillance Happens. So, how do you know you are not under electronic surveillance right now? How do you know that someone is not listening to every word you say as you sit there in your living room and communicate with your family (during the commercials)?

Now, this is not an attempt to make anyone paranoid; the chances are very slim, but again, it is a fact that Surveillance Happens. And if it happens to you, how do you know? How would you find out? If I wanted to, I could probably find a reason to visit you at your home or office and "install" one or more bugs. This is illustrated in an article that was published in *San Francisco Attorney* magazine.

Electronic Eavesdropping:
A Threat to Client Confidentiality
by M.L. Shannon

You are at your desk, reviewing documents that you will be discussing with your next client, the CEO of a South Bay electronics company. They are planning a merger with a smaller company that makes some of the components the CEO's firm uses in their products. Everything about this merger is kept confidential, as there are competitors who would benefit by having this knowledge.

The intercom buzzes—it is your secretary advising you that there is a gentleman in the reception area insisting that he has to talk to you right away, that it is a most urgent matter . . . aren't they all . . . and you still have a few minutes, so you agree to see him.

As he is shown into your office, you make the usual observations, first noticing that he seems upset, nervous . . . not at all unusual for someone consulting an attorney. . . . Before you get a chance to ask him anything, he greets you by name, explaining that an associate had recommended you.

He takes a seat, and opening his attaché case, spills some papers on the floor. Apologizing, he leans over to pick them up, then begins his story. He is an inventor who has developed an improved color multiplexed liquid crystal display, and wants to get a patent disclosure filed as soon as possible. He rambles on about his new invention, becoming more agitated every minute, and jumps up, walks over to the window, pushes the drapes to one side and mumbles something about how "they" are trying to steal his secrets.

He goes back to his seat and takes out a cigarette pack. You interrupt to inform him that smoking is not permitted in the office . . . he mumbles something about quitting and tosses the pack into the waste basket beside your desk . . . and that you are not a patent attorney. He turns and looks at you, an expression of surprise on his face, and apologizes, explaining that he thought you specialized in patent law. You escort him to the reception area, where the CEO has just arrived. "C'mon back" . . .

The details of the merger are worked out and you are instructed to begin the contract work which is expected to take about a week. The CEO rises to shake hands and, picking up his briefcase, notices a cellular phone lying on the floor, partially under your desk. He retrieves it and hands it to you. It is time for lunch, and on your way out you hand the phone to the receptionist, explaining that the inventor will probably be back to get it.

Three days later, you are called into the office of one of the senior partners. He isn't smiling. Without preamble, he informs you that Star Drives, Inc. a competitor to the South Bay company, has apparently . . . apparently, hell, obviously . . . learned of the merger and has made an offer to the little manufacturer. And the offer is being considered. The ramifications of this development remain to be seen, but for now the question is how did Star find out about the merger? Who let this out?

Remember the inventor? The only thing he invented was a good story. One that enabled him to complete his assignment—which was to bug your office. To distract you while he placed four listening devices. Four? Four.

- When he was picking up the papers he dropped, he placed a transmitter under your desk using double-sided adhesive material. (Discriminating spies prefer 3M products.)
- While opening the drapes to look out the window, he slipped a battery powered transmitter from his pocket. It had a sharp hook attached to the antenna wire and he hung it on the inside of the fabric.

- The cigarette pack he tossed into the waste basket contained another transmitter, a "throw-away" type.
- And the fourth? The cellular phone. With a few minor modifications it becomes a surveillance transmitter.

Across the street, in a parking lot, was a nondescript van. And inside it, three communications receivers, on different frequencies, with tape recorders attached. Everything said during the consultation with the CEO was recorded. And, as a backup, in the "inventor's" office was a modified answering machine recording the conversation the cellular phone was broadcasting. This guy wasn't taking any chances.

"You see, Son, most humans know little about electronic surveillance, and they have little interest in learning about it. After all, 'it could never happen to them.' That makes our job much easier."

Just that easy, sometimes. Now, read on and you will learn a lot about what you can do to protect yourself, your business, and your family. One of the first things to do is to consider probability factors. Fancy term for odds. Look for the signs. Are you in a business where the competition would dearly love to know what is being said in the board room? Are you in a relationship that you know (whether you will admit it to yourself or not) can't go on like this? Is a breakup, a divorce, inevitable? Is there likely to be a battle over custody of the children or dividing the property? Are there times when you hear someone say something that they shouldn't know? Does someone always seem to anticipate what your next move will be?

STORIES: SELF-DEFENSE

It was an intense relationship. It happened suddenly and ended suddenly, as such relationships often do. He could not deal with her infidelity and her lies, so he decided to bail out. He just wanted to be left alone to go on with his life. She was the possessive type, and "didn't want to lose him." She became bitter and vindictive when he moved out. So she tried to spread false stories about him among their mutual friends. But no one paid much attention to her, since they knew what a habitual liar she was. She called his employer and told lies to try to get him fired. They ignored her, because he had been on the job for several years and did his work well. Finally, she hired a lawyer to file a phony lawsuit for thousands of dollars, and even though it was a false claim, he had to defend himself. This was a declaration of war.

The next week he moved into an apartment building nearby and set up his listening post. He placed a modified Bearcat scanner on a window sill with the antenna facing her kitchen window. He connected a tape recorder to the scanner through a homemade relay that would turn it on whenever a signal was received. Then, knowing when she would be gone, he used his duplicate key to her apartment door to get in. In the small kitchen off the living room was an intercom used to communicate with callers at the front door of the building. It was one of the old types built into a large black box mounted on the wall. Inside was plenty of space for a transmitter, and a constant supply of power. That's where he installed the bug.

Every morning before he left for work, he checked the scanner frequency and selected a blank cassette, dated it, and placed it in the recorder. When he got home he would play it back and listen to everything that had been said in the two rooms, as well as one side of her phone conversations. Every move she made, he was always one step ahead of her. Sometimes he would arrange for her to receive misleading information through a channel that he set up; one person who knew another person . . . to see how she would react. It was driving her up the proverbial wall, but it just never occurred to her that he had placed her under electronic surveillance.

Finally, it was over. He was able to record conversations that succeed in making sure that she would never again interfere in his lifestyle. And, in spite of all her legal trickery, she never got a dime from him and ended up thousands of dollars in debt to her lawyer.

A true story. Using electronic surveillance against someone who is a threat to your freedom to live as you want may be legally wrong, but it is not always morally wrong. And declaring war on someone who knows how to bug others can be a very big mistake.

The bug was never removed. It should have been deactivated, and if there had been a way, he would have done so. However, his illegally reentering the apartment to remove or disable it was not an option. (As you will read later, once a bug is installed, you don't get it back.) It may still be there and, as it had constant power, still be transmitting the conversations of some hapless person to anyone who happens to be listening on 300.050 megahertz (MHz).

REALITY CHECK: SURVEILLANCE TV STYLE

To learn about the world of electronic spying, it helps to first know what it is not like. This is because what little most people know about it comes from what they see on TV. Of course, the amount of reality depicted on the boob tube about electronic surveillance, as with any other subject, is next to nothing, so typically, it might go something like this:

The plot has thickened, and in the second half of the "private-eye" program, a mean old man (The Bad Guy) has planted a bomb at the Snickering Sycamores Orphanage. Our Hero (the good guy) knows about the bomb. He found out from a beautiful blonde torch singer in a sprayed-on gold lamé dress at a fashionable waterfront bar and grill (one of those places frequented by guys with dark suits, sunglasses, and no necks). The only problem is that Our Hero doesn't know the exact location of the device, and there is only one way to find out: bug The Bad Guy.

Clem, sitting on the couch in the living room, glances at the clock on top of the TV set and sees that it is 10 minutes till 10. Allowing for a preview of the 10 o'clock news, eight commercials, and a station break, Our Hero has only two minutes to find and defuse the explosive device.

"It's gonna be close, Sadie."

"Clem, shut up and get me a beer."

Our Hero runs three miles through a woods full of quicksand, fights off a grizzly bear and nine security guards, and climbs over a stone wall topped with broken glass. Once on the grounds of The Bad Guy's mansion, he whips out a small crossbow, fits something you can't see too clearly to the arrow, and fires it at the terrace door of The Bad Guy's study. Perfect shot. From 300 yards away. (It has a laser sight, and you know from TV cop shows that such devices never miss.)

During the first five commercials, Our Hero has moved to the cluttered workshop of the eccentric electronic genius that some TV detective shows have. He fiddles with some of the knobs on an exotic looking device, and suddenly The Bad Guy's voice comes through the speaker—perfect crystal-clear audio. From 30 miles away. And naturally he reveals where the bomb is. (What else?)

Our Hero is airlifted in by a Marine pal who borrows a chopper. He bails out while the craft is still six feet in the air, sprints across the lawn to the nursery accompanied by frantic, pounding background music, and dashes down the aisle between rows of cribs, the wailing waifs adding to the tension. He throws open the closet door, and there it is: The Bomb.

More intense music as he hesitates, looking over the device, wire cutters appearing magically in an unsteady hand. Sweat forms on his forehead and drips down his nose. Which wire? My God, which wire? This isn't *Goldfinger*, and there is no one to show up at the last minute and calmly flip a switch to disarm it.

He grasps the yellow one and starts to apply pressure, and then at the last moment, switches to the black one . . . and . . . clips it. Suddenly, the music stops. The bomb has been defused . . . with

3.6 seconds on the digital timer. The camera pans to a scene of cheering orphans and zooms in to show several very serious looking women in black and white uniforms, clutching wooden rulers in their hands and attempting, with little success, to smile. Then, pan to close-up of grinning Hero and cut instantly to tampon commercial.

The real world of surveillance is nothing like it is portrayed on TV. Not even close. The things viewers see on the boob tube or in the movies seldom work that way. But people don't know this. Why should they? They have no interest in electronic spying. They have never read a good book on the subject. Again, why should they? After all, it could never happen to them, right? Wrong.

1
Species of Bugs

Surveillance transmitters come in many different configurations. Some of them are

- crystal-controlled or variable-frequency
- AM, FM, or digital spread spectrum (DSS)
- analog or digital
- room audio, phone conversation, or both
- powered by batteries, the phone line, or the power line
- single-stage or multistage
- high power or low power
- low frequency to microwave

So, theoretically, you could have a multistage, high-power, crystal-controlled FM transmitter that operates in the high-VHF area of the spectrum, or a single-stage variable frequency, low-power transmitter that operates on the commercial FM band. Or various combinations of the above. Each type has its advantages and disadvantages, but obviously you can't have everything in a single transmitter, so there are trade-offs. High-power transmitters have greater range but require larger batteries. Low-power are usually smaller (and the batteries a lot smaller) and so are easier to hide but will have less range. Crystal-controlled bugs are very stable and often have better audio but are usually more expensive. FM results in better sound quality, but AM gives better range for the same amount of power. Spread spectrum transmitters can be difficult to detect with countermeasures equipment but are expensive and not easy to get.

REMOTE CONTROL

As the name implies, a remote-control transmitter can be turned off and on from the listening post. The user, the spy, can turn it on and, if nothing interesting is happening (Clem and Sadie fighting again) or they hear the sounds of a search, turn it off. This means that it will not be detected with most electronic countermeasures equipment, excepting possibly the nonlinear junction detector (NLJD). This is a complex and expensive device that detects solid state components (chips, transistors, diodes) by emitting a radio wave that is reflected back to the device.

Remote Microphone

The idea here is that the microphone and the transmitter do not have to be in the same place. The reason for such a setup, as should be obvious, is that the signal may not be detected during the RF portion of a sweep. The microphone will be found in the physical search if it is done right but might not be if done by amateurs.

Now, this isn't quite as simple as it sounds— stringing a wire from the microphone to the transmitter because the wire may pick up interference that will make conversations difficult to understand. Also, the longer the wire, the higher the signal loss, and by the time the audio gets to the transmitter it may be too weak to provide recognizable speech. There are audio preamplifier stages that can be interfaced to the transmitter, but an easier, if more expensive, solution comes from C Systems. They market the HWA1100, which is a multi-channel system that has 10 audio channels and 1 phone line channel available, operates from the power lines, and has a built-in nicad backup battery. With a two-stage amplifier, it is possible to use cables up to one mile in length; the microphone can be that far from the transmitter.

Top: This is the gizmo described in *The Phone Book* that a neighbor thought was a bug. It is a filter used to keep radio stations from interfering with the phone line. However, it could have been a series line-powered phone transmitter— PK Elektronik makes them as small as a pea, and one of them could have been used on this circuit board.
Bottom: Researching a book can be expensive. This is an old Deco crystal-controlled UHF transmitter that I burned out while testing. Increasing the battery voltage slightly may increase power output and range, but be sure you know what you are doing. Deco no longer produces audio transmitters.

A fascinating system, methinks. Use as is or wire the transmitter into the listening unit at a distance. Use a contact microphone disguised as a glass break detector and you have one helluva setup to intercept conversations. However, if you try to import it, U.S. Customs may well intercept you.

CRYSTAL CONTROLLED

A crystal-controlled transmitter uses a quartz crystal (called a "rock") to determine its operating frequency. Quartz is piezoelectric, meaning that if a small voltage is applied to a thin slice of the mineral, it will vibrate (oscillate) at a very precise, constant, rate. This makes the transmitter very stable. It will stay on frequency without drifting, even when handled, moved around, placed near large metal objects, and so on. The only way to change the frequency is to replace the crystal with a different one. Most crystal-controlled types transmit in the high VHF or low UHF area, so you cannot receive them on a standard FM radio—you need to use a scanner or a communications receiver. Crystal-controlled models are usually more expensive, as stated above, and for some reason seem to have better audio quality.

VARIABLE FREQUENCY

An analog variable-frequency transmitter uses a tuned circuit to generate the frequency. A tuned circuit consists of a coil of wire and a capacitor. The advantage of the variable-frequency types is that the user may be able to adjust the frequency to fit the situation. This can be important for a number of reasons, including the selection of an antenna, as you will see later. One problem, however, can be stability. The very cheap models, such as the single-stage (one transistor) modulated oscillator, will drift if they are moved around, and also as a result of variations in ambient temperature and in supply voltage and current draw. This, the amount of current, varies; when it is transmitting audio it draws more current, and when no sound is present to be transmitted it draws less. This change in current can have a slight detuning effect, since the one transistor is doing the job of oscillator, amplifier, and modulator. A better transmitter will have four or more transistors: an audio preamp, audio amplifier, oscillator, and RF amplifier. One of the best variable-frequency models was the old VT-75. It is no longer being manufactured but may be available from some underground suppliers and so is used as an example in this book. The VT-75 stayed on frequency very well, even when used as a body wire. Only occasional minor retuning of the receiver was necessary. I have tested the VT-75 in a variety of situations, and performance was very good. But, all other things being equal (which they seldom are), a crystal-controlled transmitter is the best choice for stability.

THE BURST TRANSMITTER

This rather exotic device picks up sound through a microphone, like any other bug, then converts it to digital and stores it in one or more chips. The principle is the same as "tapeless" telephone answering machines. At certain intervals, the stored data (conversation) is transmitted in a short "burst" that lasts only fractions of a second. How often these bursts are transmitted depends on how much sound the device can store, i.e., how much "memory" it has. The more memory available, the longer between transmissions, but the trade-off is an increase in size (due to more chips) and higher power requirements. This means larger batteries which are easier to find in the physical search.

They are quite rare; one of the best sweep teams in The Biz tells me that it has never encountered a burst transmitter in its searches.

REPEATERS

A repeater is a device that takes a signal from a low-powered transmitter and rebroadcasts or repeats it at a higher power level, which increases the effective range of the bug. The principle is the same as the repeaters used for two-way mobile radio communications. So, one could install a very tiny transmitter, perhaps a phone line type the size of a pea, which normally has a range of less than 100 feet, and, using the repeater, locate the listening post a mile or more from the bugged site. WAID has a nifty little repeater, the model BXR-2202, which goes for only four grand. Size is 3 x 3 x 13 inches, it has 2 watts of power, and it works in the 150 to 174 MHz area. The battery life is up to 10 days, depending on use. Just think of all the places you could bug with a 20 MW transmitter and one of these! ICOM America also makes repeaters, intended for hand-held two-way radios, that can be interfaced to a bug, and for a bit less cash than you'd have to fork out to WAID.

Most of these transmitters and repeaters can be had if one has the cash. PK Elektronik has them, and while you can't buy them in the United States, they are available in Germany through the PK retail division. They are also legal to sell in some other countries—Ireland, for one, where Sheffield Electronics may be relocating its manufacturing facility. Just remember that you will probably be violating customs laws if you bring them back to the United States.

MICROWAVE

Microwave is generally considered to be that portion of the RF spectrum above 1,000 MHz, but I don't know that there is any "official" definition. As far as surveillance is concerned, there is no sharp dividing line, a point at which the characteristics of bugs is in a different category from those of lower frequency. There are indeed differences, as you will read later in this book. For the most part, microwave surveillance is complicated. The equipment is often much larger than for lower frequencies, is very expensive, and requires antennas that are not appropriate in many surveillance installations and are not readily available in the form that is useful for ordinary RF "down-and-dirty" surveillance.

Put another way, a microwave surveillance transmitter is either going to be very expensive or will require some knowledge and experience in electronics. If you have the bucks, there are commercial units available, such as the PK Elektronik model 560 Light Bulb Transmitter, built into a nonworking light bulb and having a range of 100 m, and the model 570, which is built into an ash tray and has a mercury switch that shuts it off if it is turned upside down. (Why would anyone turn an ash tray upside down except to empty it?) It claims a range of 250 m and a battery life of 100 hours. Very nice stuff, I am told, and very expensive.

If you want to take a crack at building your own, Innotek Inc. used to sell a small (2.25 x 3.75 x 1 inch) X band transmitter kit for $29 that included instructions on how to modify it to transmit audio. The last address I have is:

P.O. Box 80096
Fort Wayne, IN 46898

OK, you ask, what does one use for a receiver? A modified X band radar detector. In the February 1991 issue of *73* magazine (a publication for amateur radio operators) is an article by Steve Noll that details how he modified a BEL XKR radar detector to receive this frequency range. According to the author, "At a convenient location near my house, about ten miles from the beacon, I can get a solid signal." This was using the horn, the small antenna that comes with the radar detector. He doesn't say what the power output of the beacon is. The modified receiver was built into a 7 x 4 x 3 inch plastic box.

This project is not for beginners; it requires experience and sophisticated test equipment. However, a spy who is worth his salt will know someone who can put this system together. For a price.

Amateur radio operators can be seen as well as heard—on Ham TV. Hundreds of licensed operators have built their own TV stations using the many products available on the market. This company (http://www.hamtv.com/cat97.html) has the TXA5-RC transmitter board with 1.5 watts power for $129. It is video only but the company also sells an audio board that can be incorporated into the transmitter. Log on for the latest information.

The Ultimate Bug?

Suppose you had a transmitter that was not made for nor intended as a surveillance device, and therefore apparently legal to sell. It could be easily programmed with a computer (virtually any computer) to transmit on any frequency from 130 to 150 MHz. Not only that, it could be set up to change frequencies, moving (hopping) from one to the next to the next, over and over, every few milliseconds. And has a variable power output as high as a full watt; a very powerful bug. Such a device exists; the Micro PLL transmitter manufactured by Agrelo Engineering in New York. It used to sell for $285, which is reasonable for all that it can do, and there is nothing else available on the open market that even comes close at any price. Their Web site is www.agrelo.com.

MODULATION TYPES

Without modulation, a radio signal is nothing more than an unwavering tone as heard on a communications receiver. Modulation is the process of doing something to the signal so that it can transmit information. There are various types of modulation, such as the AM and FM that we all have heard of; some that are not of interest here, such as pulse code; and two other types that can be used in surveillance transmitters: single sideband (SSB) and DSS.

Frequency Modulation

FM has been the standard for many years, and while DSS and SSB are being used in the exotic and expensive government transmitters, the majority of bugs are FM. The audio to be transmitted causes the signal to vary in width (called "deviation") according to the frequency of the sound. For example, two-way radio systems have a bandwidth of plus and minus 5,000 cycles, so the audio response, or quality, is limited to this range. Good enough for voice communications, but not for music. FM broadcasting has a wider signal, so that audio of 20,000 cycles, the range of human hearing, can be transmitted. This is why commercial FM stations are spaced farther apart on the band than AM and two-way radios.

Amplitude Modulation

An AM signal doesn't vary the bandwidth of the signal, it changes the amplitude, or "strength." Ask someone in The Biz what AM is, and they may respond, "Ancient Modulation." Why? Because virtually all of the surveillance transmitters made in the last 40 years are FM. The reason for this has to do with the early development of radio, and AM is said to be "obsolete" for surveillance use. People in The Biz scoff at the mere mention of an AM bug. The difference between AM and FM in surveillance is the same as it is in commercial broadcasting. AM is prone to fading in and out (called "QSB" by radio operators) and is more subject to interference. The clicking and whining you hear on an AM car radio while in traffic is one example. FM is usually heard more clearly: either well or not at all.

One old pro that I was talking to laughed and said that no one in their right mind would use an AM bug. "Why, that's nonsense," he said. "AM is never used for bugs. It is too prone to interference, and it 'just isn't done.'" While it is true that AM has its problems, AM transmitters do indeed work, as described in a true story in *Don't Bug Me*.

Some thoughts on AM:

- An AM transmitter requires more supply power to operate than FM of a comparable output but has greater range.
- At very low signal levels, an AM signal can be received better than FM.
- A very high-power AM transmitter can (still) be built from an old tube-type radio. The output can easily be several watts, and with the right antenna, range can easily be several miles. (This is what I used in my short-lived career as a pirate radio disk jockey—until the FCC came out and made me shut it down.)
- With a certain amount of tweaking, the frequency can be shifted to below the standard AM band (lower than 550 kilohertz [KHz]), making it more difficult to detect. The plans for such a transmitter were published in *Popular Electronics* in the early sixties. I don't know if a reprint is available.
- The cost of modifying the radio into a transmitter, once dirt cheap, is still affordable, as tube-type radios can sometimes be found in second hand stores and "junk" shops. A few bucks and an evening's work and you have a powerful bug.

Single Sideband

The details of SSB are complex and too technical for this book, but it is essentially a modified form of AM. Only one of the two "sidebands" that contain the information is transmitted, and the carrier, which modulation sort of "rides on," is suppressed or not transmitted. It is "regenerated" by the receiver. The advantage of SSB is that the signal packs more punch than AM. A lower-power signal can be received at a greater distance. Another advantage, depending on which side of the fence you are on, is that SSB cannot be received—it cannot be understood—by an ordinary radio. It sounds a little like a mechanical Donald Duck. So, the signal is less likely to be intercepted by accident. A little more secure. A few years ago, back when there was an "open" market, some high-quality SSB transmitters were being imported from the USSR, allegedly made by the KGB. (In the USSR, everything is blamed on the KGB. In America, everything is blamed on the CIA. Both are guilty.)

I do not know of any SSB bugs available in the United States. If you search, you may find.

Digital Spread Spectrum

Spread spectrum has a fascinating history. As I have heard the story, it was first conceived of by film actress Hedy Lamarr. While living in Europe in the years preceding WWII, Hedy was married to one Fritz Mandl, who was a dealer in munitions. Mandl was selling his wares to Herr Hitler, which Hedy didn't like very much. She knew about her husband's business affairs and picked up a certain amount of information, which is how she learned about a problem with controlling torpedoes. It seems that they were steered by radio signals, but these transmissions were easily jammed by the enemy, causing the torpedoes to malfunction—to miss the targets or detonate before reaching them. So Hedy came up with the idea of frequency hopping: the control signal information was not transmitted on a single channel but rather jumped, or "hopped" from one to another, with each hop carrying small bits of the signal. The only way to reassemble the data was with a receiver that was synchronized with the transmitter. This was originally done with spools of punched paper tape similar to those once used with teletype machines.

Hedy immigrated to the United States and, with the help of several engineers, was able to make the idea work. She patented it in the early forties. It was apparently used by the military forces in those years of WWII, though not to any great extent, and was declassified in about 1985. Today, the government and corporate America use DSS, but it seems that no one has paid Hedy a single dime for her invention.

How Spread Spectrum Works

Spread-spectrum technology is complicated and difficult to understand, so my explanation here is very basic and oversimplified. Spread spectrum "spreads" a radio signal over a much larger portion of the radio spectrum than is actually needed to propagate the signal; it increases the bandwidth of the signal. As you read above, bandwidth refers to how much of the radio frequency spectrum a particular signal occupies. Commercial television stations have a bandwidth of 6 million cycles, or 6 MHz. More than a thousand two-way radio channels would fit into the space wasted by one television station, which would be a much more productive use of the very limited spectrum. (Yes, it's true, I badmouth TV every chance I get!)

Because DSS signals are so wide, they transmit at a much lower "density" than narrowband transmitters. For lack of a better analogy, think of FM as filling a bucket (radio) with the water (signal) coming out of a tap and spread spectrum as trying to fill the same bucket (radio) with the same water, but now it is falling from the sky like rain. It is the same amount of water (signal), but is spread over such a large area that the bucket (radio) would catch only a tiny fraction of said water (signal). To get all of the water, the bucket would have to be hundreds of feet in diameter. Continuing the analogy, the receiver would have to be able to understand transmissions many times wider than an FM radio. So, the basic definition of DSS is that it produces a signal that is much wider than is actually necessary to transmit the information.

This is done by taking the output of a pseudorandom noise generator (called the "spreading code") and mixing it in with the signal to be transmitted. This same code is used in the receiver to convert the signal back to its original size so it can be understood.

If you have a scanner or communications receiver that tunes the FM broadcast band, set it on narrow band FM (NBFM) and tune to a local station. You will recognize the sound—commercials and occasional music—but it will be "distorted" or "clipped," because you are hearing only part (a "slice") of the transmission. Now, if it were possible to set the scanner, which in NBFM mode is 5 KHz, to one-tenth of that or 500 cycles, you wouldn't be able to recognize anything being transmitted because the FM signal would be so wide that very little of it could get through. By comparison, commercial DSS systems use bandwidths of up to 100 times the information rates required, and military systems may use spectrum widths from 1,000 to 1 million times the information bandwidth. At such enormous bandwidths, most bug detectors wouldn't be able to hear the "noise." They might not even know that the signal was there unless they were right on top of the transmitter.

Spread Spectrum Advantages

In addition to the security (LPI, or low probability of intercept) and resistance to jamming, spread spectrum has other advantages. It has greater penetrating power and isn't as affected by walls, ceilings, and other obstructions in the signal path compared to AM and FM. And it also has much greater range. Ergo, a DSS bug, using the same amount of power as its NBFM counterpart, is capable of establishing successful surveillance over a distance of miles, literally, compared to hundreds of feet.

....

> While doing the final editing, I took a break and was discussing my bill with Pac Bell. I was advised that I could purchase (on the easy payment plan) a cordless phone from them.
> Ho humm . . .
> Not just any cordless phone, I was informed; this is a 900 MHz digital spread spectrum (DSS) model. And it has all sorts of custom calling features, including Caller ID.
> Ho humm.
> Then, I am told that it has a range of "up to a mile and a half." Hmmm. With some major modifications, a person could build a really snazzy bug out of one of these.
> Reminded me of a rumor that was going around a while back (don't remember the source or I would list it here) that some manufacturers of DSS cordless phones

compromised their resistance to eavesdropping at the request of a certain federal agency. I can't say it's name, but the initials are National Security Agency. I have no idea if this is or is not true. On one hand, the feds are not above such things in their war on privacy, and on the other, there are a great many "urban legends" floating around the Internet, which are 99-percent bullshit. I neither believe nor disbelieve this, but regardless, never forget that any time you use a telephone of any kind, there is always the possibility of your conversation's being intercepted.

Spread spectrum also provides an answer to the problem of the crowded radio spectrum. Back in the early days of radio, the spectrum was virtually unused. There was international shortwave broadcasting on the high-frequency bands, but very high frequency, or VHF (30 to 300 MHz) was barely touched, and UHF (300 to 3,000 MHz) hadn't even been explored save for radar. As technology improved, it became possible to operate radios on frequencies that were not feasible a few years previous. But, as the use of radios exploded, it wasn't long before the spectrum became crowded. Now, in the nineties, virtually all frequencies, all the way up to where radio waves become infrared light, have been assigned. Spread spectrum eliminates this problem, at least for some radio services. Thousands of them—police, medical emergency, law enforcement, business, cellular telephone—can operate within a range of frequencies, where with conventional NBFM only hundreds would be possible.

Spread Spectrum Types

There are two main spread spectrum types: frequency hopping (FH) and direct sequence (DS). Frequency hopping is what the name implies, although an oversimplification: the signal hops around on different frequencies within its bandwidth. It is possible to intercept FH with certain types of receivers, but depending on the specifics of the transmitter, you might hear recognizable speech and you might get your ear bent by a series of clicks or pops or tones.

Frequency hopping is supposedly more difficult, if even possible, to intercept and understand, and more difficult to detect, although not impossible with the right equipment. This depends on the bandwidth and other factors too technical to get into here, and which I really don't understand all that well.

Detecting Spread Spectrum

So, how would you go about finding a DSS bug?

DSS will show on a spectrum analyzer, but the operator has to know what he's looking for. Spectrum analyzers aren't cheap; the good ones are very expensive. One of the countermeasures receivers from Marty Kaiser Electronics will not demodulate the signal, but it will detect the presence of a DSS transmitter if it is within a short distance.

Dynamic Sciences International markets the SSIU-1 Spread Spectrum Identification/Detection Unit, which, according to the company's Web site, "works in conjunction with the R-110B Wide Range Receiver to detect and characterize a variety of spread spectrum and digital communication emanations at negative received signal-to-noise ratios." The SSIU-1 utilizes a proprietary analog signal processor to perform the detection function, and the outputs of the unit may be displayed with any RF spectrum analyzer. According to the Web site, "The system has two basic modes: direct sequence (DS) and frequency hop (FH). Detection consists of viewing the spectrum analyzer and recognizing the distinctive waveform structures produced by the SSIU-1 when (and only when) a target signal is present. Spread spectrum, symbol rate, and hop rate are also easily and quickly determined from the spectrum analyzer presentation. The approach takes advantage of the human operator's inherent capacity for accurately assessing the character of the processed waveform, permitting identification of emanations without triggering false alarms." In other words, knowing what the hell they are doing. (Details are at http://www.dynamic-sciences.com/eands.html.)

Spread Spectrum Bugs

The SIM DSS 2000 is a remote-controlled encrypted spread spectrum surveillance system. The power output is 400 M, and the signal is spread up to 10 MHz. It is made in Germany and is not, to my knowledge, available in the United States. Know anyone in Deutschland?

There is a "pseudo" spread spectrum body wire transmitter made by Textron in Hanover, Maryland. Pseudo (their term) means that it is not actually spread spectrum but resembles DSS in that the signal is very wide, something like one megacycle. This is a digital transmitter that is about 2 1/2 inches square and half an inch thick. The high-power model puts out 750 mw and has a range of several blocks, depending on conditions. The receiver has a signal strength meter to measure the transmission and two "Vu"-type meters (they are stereo transmitters) that measure the strength of the audio being transmitted and not the volume from the receiver. This means that at a glance the user can see if the transmitter is sending anything, if there is interference in the area where the transmitter is located (such as loud background noise), or any other problems. They have two antenna connectors that work together for two different antenna types. So if there were to be multipath distortion (see glossary) due to the location of the transmitter changing and the signal being reflected, one of the two antennas would still be able to pick up the transmission. The output of the receiver can go directly to a digital audio tape recorder and even to a CD-ROM programming system.

A fascinating system, but a little spendy for most of us, as it sells for about $8,000.

American Microsystems, Inc. (AMI) makes a number of DSS chips, which are used mainly with a computer, for data transmission—a wireless modem. However, they can be used as a DSS voice transmitter. This requires that the output of a small audio amplifier feeds into an analog to digital converter (ADC), possibly preceded by a splatter filter (clipper) and then into the transmitter chip. An electronic technician should be able to do this, but it is not a project for beginners. Who do you know that would build you such a system?

Pop Quiz
One of the AMI chips, The SX-041, requires 3.9 volts at 10 milliampere (ma). How long would it transmit using two "D" cells for power? You should be able to figure this out from the information presented later in this book. Hint: read about cutoff voltage.

BUGS IN DISGUISE

Having had a look at types of transmitters, let's look at a few of the many ways they can be packaged. A few years ago, there was a discussion on one of the computer BBSs about old, broken, obsolete electronic and computer stuff. People were talking about their basements and garages being full of junk they will never use, and the idea of a massive collective garage sale was considered. It never came to pass, so I posted a message asking people to please not throw anything away (one person's junk is another person's treasure, and the like). In response to that ad, several months later, some anonymous person sent me a large box of "junk." Among the items were several old nonworking body wires made by Spectrum Communications. They are similar to the type you see used on the older cop shows: The captain is using adhesive tape to stick the transmitter to the undercover agent's chest (got to be a bitch to remove) and stringing the antenna wire up his coat sleeve, wishing him luck, and all that. Invariably the bad guys discover that the agent is wired and are just about to shoot him when the rest of the cops kick the door in and shoot all the bad guys. TV. Ho hum . . .

These particular units are a little larger than a king-size cigarette pack, are built into an expensive machined aluminum case, use expensive Lemo connectors, and operate from an internal 12-volt rechargeable battery pack. Very nice stuff—they have a power output of about a quarter watt, and the metal case acts as the base of a ground plane antenna. None of them worked, but after replacing a 741 op amp chip in one of them and injecting audio into the second stage, I fired it up. It transmitted on 148.005, just outside the amateur two-meter band, and so I could receive it with my old Kenwood. The sound quality was excellent, with the audio not being muffled or sounding distorted (I used an FM radio playing Schubert), even though it was designed for a body wire rather than a room audio transmitter.

An interesting illustration of how things were, not so many years ago. Today, body wire transmitters are available in digital stereo. Like the Textron unit mentioned above. But Textron doesn't sell to the general public and do not want contact information published in this book.

Who do you know that could find them and buy one for you?

In issue 24 of *Full Disclosure* magazine (unfortunately out of print) are pictures of some of the surveillance equipment that was on display at the National Technical Investigators Association (NATIA) conference in Washington, DC. One is of an ordinary disposable cigarette lighter, into which is built a short-range transmitter. One of the new generation of off-the-body wires. On second thought, at $1,150, it isn't so disposable.

Transmitters are built into beer cans, briefcases, umbrellas, tape recorders, calculators, stuffed animals, TV remote control units, pepper shakers, and I even heard of one built into a mouse. A computer mouse. They're being built into damn near anything you can think of but might not have. So, if you are sitting in a bar some night and some stranger is asking you a lot of questions, look for the signs. If he keeps turning his lighter so it always has the same side pointed at you while playing eight ball, has a cigarette pack in his pocket but never smokes, or will not take his hat off, be a tad suspicious. Surveillance bugs are like their biological counterparts; you never know where the li'l critters might be hiding. Keep that in the back of your mind as you read on. There's another pop quiz coming up, and you will be expected to pass.

WHERE TO GET TRANSMITTERS

Ah, the good old days, when the first edition of *The Bug Book* was written, where surveillance transmitters were readily available to anyone who wanted them. You could walk into any of a dozen "spy" shops and purchase any number of bugs. Some were very good; many were very mediocre. Most were not capable of what the sellers or advertisers claimed, and virtually all were overpriced. More often than not, you didn't really know what you were getting, or how well it worked. It was common to hear hype like, "As small as a dime, will pick up a whisper 50 feet away, transmit it 10 miles, and run for weeks on one little calculator battery."

Bullshit.

In a *New York Times* newspaper article, a spy shop owner claimed he has a bug that is the size of a pinhead. I think the price was $35,000. Something like that. It was said to have a range of "miles."

Bullshit.

And how could you tell one from another? Ask for a demonstration. If what they were selling was for real, they would be happy to do so. But if they were lying, they would come up with any number of excuses. "Uh, well, that particular model is sold out right now. Can't get 'em fast enough for our customers. Yep, got 10,000 back ordered. However we do have in stock this little gem . . . not quite as good, but . . ."

Bullshit.

But, the point is that they were available. And if you knew what to look for, how to tell the good stuff from the poor, you could usually find something that would get the job done.

Thanks to the feds, this is no longer true. The few retail "spy" stores left either don't sell transmitters, or have low-end types in kit form. There are some available from retailers that don't have the word "spy" in their name (this is a red flag for the feds), which are listed in Appendix C: Places to Get Things. Also, a few mail order sources of some good equipment are listed, but availability is not guaranteed. Sources come and sources go, but merchandising is merchandising. And bullshit is bullshit.

Weasel Willie and the Black Market

OK, your contact has informed you that Weasel Willie has some stuff in his garage over near 145th and Division. So, you go over to visit him with a money belt full of cash, but how can you tell one bug from another? The good ones from the mediocre? How do you know what you are getting on the black market? Well, first of all, not necessarily from the price. Not then, not now. Some of the old "spy" shops obtained low-cost imported transmitters for about $30 or so and resold them for 10 to 20 times that much. Or more. The Spy Factory chain was well known for doing this. So how do you know that Weasel isn't going to rip you off? Ask about the power output, range, whether it has an antenna matching adjustment, and whether it has a separate audio preamplifier, or pre-emphasis stage. If possible, try before you buy. Insist on a demonstration. If they will not demo it, go somewhere else. (If you can find somewhere else.) You might also take a copy of this book with you. Rip-off artists will shrink away from it like vampires from the rising sun.

The Internet

I spent a few hours searching and came up with dozens of sites that offer surveillance transmitters. Much of what I saw is as described above: bullshit. Bugs that are advertised as having a power of 20 mw and a range of "1,700 yards," or just short of one mile. The usual hype. This is not to say that all such sites offer overpriced junk, but it does make me wonder about something. Since most of the retail "spy" shops went out of business following the U.S. Customs raids on the Spy Factory chain, why are these Internet companies still operating? I don't recall seeing a disclaimer advising that their products are sold only to law enforcement agencies. Give it a try if you like, but remember that Big Brother is watching.

Wireless Microphones

If you purchase a brand name wireless microphone system, there is one thing you can be Shure™ of: you will get what you are paying for. Very high quality. Excellent equipment. But again, we have a trade-off. You will be paying plenty. These systems run from the low hundreds to well into the thousands. Sennheiser. Sony. Shure. Many wireless microphones are suitable for surveillance, at least short range, although this is not what they were designed for. Some of the transmitters are too large to conceal easily; others are not. And the receivers, which make up much of the price, are not necessary for surveillance work. They are made to receive the signal from the transmitter, but a communications receiver is probably more sensitive.

You already knew that, didn't you?

One example is the Shure™ LX Series. Transmitting frequency can be from 169.445 to 240.000 MHz with an advertised range of 91.5 m (300 ft.) under typical conditions with 50 MW output. Other models are advertised at up to 600 feet, which is more than enough for most surveillance applications. If interested, see their Web site: www.shure.com/publications.html#order.

Keep in mind, though, that the advertised range of wireless microphones is with the attached antenna. Use a better one, with forward gain and in a higher location, and the range increases. This is not exactly legal, as the FCC type acceptance no longer applies if the device has been modified, but anyone risking the consequences of bugging their neighbor isn't likely to be too concerned about that.

The Shure EC1 Transmitter Specs

RF power output: 50 MW max
Modulation FM ±15 KHz deviation, 62 microsec pre-emphasis
Battery type: 9-volt alkaline (Duracell 1604 or equivalent)
Battery life: 7 hours minimum (alkaline)
8.4-volt nicad optional
Current drain: 55 ma typical
Antenna: attached, 15.2 inches (386 mm), flexible wire
Dimensions: 3 1/2 x 2 1/2 x 1 1/32 inches
Weight with battery: 5 ounces (oz.)

Remember there's a pop quiz coming up. You can use these figures from the Shure EC1 in your calculations of battery life and see if you get the same results. Hint: read about cutoff voltage.

And where can you purchase these expensive goodies? Not from spy shops or Web sites with names like Spy World or We Spy 4 U or whatever. Not unless you want to pay the considerable markup they charge. There is a better way. A large book that has sources of these microphones as well as other equipment that can be used for surveillance. A book you already have. Yep, you guessed it: the Yellow Pages. Look under TV Stage Equipment, Microphones, and also, buried between Attorneys and Automobile, you will find Audio-Visual. You may have to spend a lot of time talking to receptionists that have no idea what you are trying to ask them, but keep at it and you will eventually get connected to a technician who loves to talk about his job. Amazing what you can learn if you persevere. You just might find a place that has old "obsolete" equipment gathering dust and buy it for the proverbial song. A little social engineering might help. Make up a few flyers advertising a new musical comedy opening in a few weeks at one of the local "very off Broadway" theaters. Go in to the shop and explain that the show is running on a shoestring and you would appreciate a break if possible. Add that you will see to it that they are listed in the program as having helped make it possible.

Another possible source is second-hand stores and pawn shops. Last week, I saw what looked to be a good quality wireless microphone in a hoc shop for $79.99. I don't remember the exact frequency; it was about 170 MHz. If worse comes to worst, you might settle for the units that one of the chain stores sell. I think you know who I mean. Some of the stuff they sell is good, but a lot of it isn't, and I do not often recommend them for much of anything.

Space-Age Hearing Aids

Other products that were never intended to be used as surveillance devices, but can be, may operate on any of a number of frequencies. Available from hearing aid stores and organizations for the hearing impaired are transmitter-receiver sets that operate on infrared as well as FM. They are supplied so that they can be attached to a stereo or TV set—or a microphone in the case of some models. One manufacturer is Phonic Ear Inc. They're expensive; the infrared models start at $229 and the FM at $535, but the quality is very good.

IMPROVISE

Cordless phones, operating on 46 and 49 MHz, can be modified and used as bugs. Start by removing the case and, depending on your experience with electronics, whatever else that isn't necessary, to make it as small as possible and therefore easier to hide.

One can also build a bug from the single chip that cordless phones use. Only a dozen or so other components are needed, and the power output can be up to 100 mw. And, with the linear amplifier in Appendix C: Places to Get Things, this can be increased to a full watt.

The schematic diagram is in the Motorola Telecommunications Device data book, DL136. It used to be free, probably still is, and is available from:

Motorola Literature Distribution Center
P.O. Box 20912
Phoenix, AZ 85036

Look, Ma—No Hands

Cordless headsets for hands-free telephoning are available from various sources, including Hello Direct, at www.hello-direct.com/scripts/hellodir.exe.

One model, a Panasonic, operates on 2.4 GHz which eliminates monitoring by the average scanner user, as scanners don't tune that high. Others are on the 900 band (902–928 MHz), which many scanners do cover. These are short-range systems, but, as with any other transmitter, the distance at which they can be received can be increased by using a good communications receiver. Interesting what you might hear from a vehicle parked outside the buildings of businesses that use these systems. Especially if you know, personally, someone who is employed there.

Two-Way Radios as Bugs: Not!

Some of the "spy" shops sell hand-held two-way radios, usually in pairs, that operate in the 450 MHz UHF band. They have a power output of one to five watts, which is good for up to a mile or so range, depending on conditions, of course. Some sales reps may tell you that these can be used for surveillance. Not likely. In this frequency area, they are very likely to be overheard by others. Also, this type of radio was not made for continuous operation. It is possible that they could overheat and burn out if left on for extended periods of time. And, at such a high power output, the batteries last for only a few hours if that.

TONE SQUELCH AND OFFSET

Something else I heard at the old Spy Factory in San Francisco was that the 450 MHz area really is safe for surveillance transmitters because the services that operate there use "tone squelch" and "offset" so others will not be able to hear the bug. I will explain.

Most radio services in the UHF area are on repeater systems, and because there are so many businesses that have radios and due to the limited number of frequencies available, often several companies share the same one. In order for one business to not have to listen to the transmissions of the others, they use tone squelch, such as the Motorola Private Line or "PL." When one business uses the system its transmitter sends out an audio (PL) tone that is used to turn the repeater on, to "key" it. The repeater then sends out the same tone to the receivers in the mobile units, and when they hear that tone, they "turn on" and can hear the signal. Each company has a different audio PL tone so they hear only the transmissions intended for them. That way Hymie's Taxicab company will not hear the dispatcher of Tree Frog Trucking Company, and vice versa. Now this does not mean that one can't hear the other. Most radios have a monitor button. Pushing it will bypass this feature and allow you to hear everything being transmitted on your frequency—including, if you are within range, a bug.

The same thing is true of some types of voice pagers. An interesting example: I was sitting in a cab at an airport, waiting for an associate to come out of the terminal, when the driver's beeper went off. The dispatcher was asking one of the other drivers if he wanted a particular fare; a very expensive fare to a distant city.

The driver of the cab I was in wasn't supposed to hear this. He was openly very unhappy, and when I asked what was the deal, he explained. The dispatcher was known to give a lot of the better rides to this particular driver, named Louie, because he was the "owner's pet." But, the owner wasn't aware that the pagers they had recently issued to the drivers had this feature, and that the drivers who knew how to use it could hear pages to all of the drivers. My driver called in and started raising hell, followed by some of the other drivers. I suspected that 'Louie' wouldn't be the fair-haired boy much longer.

The term offset means only that the transmitters and receivers do not operate on the same frequency. Suppose the Tree Frog Trucking Co. is assigned the frequency 461.450. The repeater output will be on that frequency. The mobile units transmit exactly 5 MHz higher, in this example, 466.450. The signal from the mobile units goes into the repeater at 466.450, uses the PL tone to activate it, and then the signal is rebroadcast at a higher power level on 461.450. This, 461.450, is the frequency the mobile units receive on and the one you want to punch into a scanner to hear both sides of the conversation. So offset has nothing to do with whether or not a surveillance transmitter will or will not be heard.

THE CELLULAR BUG

Remember the "inventor"? The idea of using a cellular phone as a surveillance transmitter occurred to me about six years ago while I was working as a technician for an electronics company that manufactured cellular telephone interception systems. I was testing a new device we had just started production on, which was based on the Oki 900 phone. I named it "Cell-Scan." With this system, the Oki can be used as a normal cellular phone, but when in use as a monitoring system the power switch is placed in the OFF position, and so the display is blank. The phone is controlled, through the cable, by the computer.

So, I wondered, why can't the same thing be done without it being part of the Cell-Scan system? Why not modify it so it can transmit with the power switch and display turned off? Just bypass the power switch and replace it with another, hidden under the battery, along with a second one to turn the display off and on. A fascinating idea, since it would eliminate the two main problems in installing a room audio transmitter. The first is range. Without the chance to test a surveillance transmitter, it's hard to be certain that it will send its signal to the listening post. But cellular phones have no such limitations. They will transmit, through the cellular radio system, to any phone anywhere in the world.

The other problem is installation. Instead of finding a place for the bug, find a reason to get into the target area, call a prearranged number, and just leave the phone there. Maybe hide it under a desk or whatever. Right, like the "inventor" did. And to eliminate the possibility of the phone being traced, it can be rented on a prepaid basis. Big deposit, no ID required, very high per-minute rate. An excellent idea for short-term surveillance. The new lithium batteries are good for five or six hours. You will test it first, won't you?

A Better Way

To overcome the problem of the battery running down, the obvious solution is to use a larger battery. Now, you can't very well leave behind a cell phone with a dozen C cells taped to it (it would most assuredly attract attention), so let us look at a better way. Suppose we were to take a cell phone and reduce it to its smallest possible size. Remove the case, as it isn't necessary, and we could eliminate the display if it were already programmed. Then we get an old transportable "bag" phone and do some major surgery. We remove the circuit boards and place the modified small phone inside, along with as many lithium batteries we can cram in. Voila! We have a working cell phone, with many hours—maybe even days—of battery life. And, as the display on the handset of the transportable does not work, it is assumed that the phone is turned off or the battery is discharged. Dead. Harmless.

The Old Bug in the Battery Trick

What with people getting wise to many of the spy tricks of contemporary corporate espionage, they might be a tad suspicious if someone left a cell phone behind. So, they might think they can defeat such a trick by removing the battery. "If it is a 'bug,' it can't transmit now, can it?" they reason with a certain smugness. Well, not necessarily. It wouldn't be difficult for the average technician to build a small bug into a cell phone battery. And what with them getting very cheap (a free one when you fill your gas tank?), they become "disposable" bugs. A great idea, but one I can not take credit for. WAID, so I hear, is already manufacturing them. These people are really on top of things in the world of surveillance.

No Disassembly Required

Another possibility. Cellular phones are controlled by a microprocessor, the same as a computer. So, there is an existing program, in a chip, usually an EEROM, inside that contains the instructions that make the phone do what it does. It is possible to rewrite these instructions to make the phone do things is wasn't designed to do. Cell-Scan, for example, only instead of switching the mobile ID number (the number dialed to call the cell phone), it would be set to listen for a certain code sent from the eavesdropper. When it was heard, the ringer would be silenced, the display turned off, and the phone would auto-answer. So if you were driving down a freeway and someone activated such a modified phone, they would be able to hear what was being said in the vehicle. Theoretically, this could be done without taking the phone apart. The Motorola models, perhaps, could be modified by plugging a palmtop computer into the service jack and transferring the program if someone had temporary possession of your phone. Perhaps while your car was being serviced or repaired. *Nuts & Volts* magazine ran a series of articles on cell phones, which are most interesting, and reprints are available.

Another Way

Some phones already have an auto-answer feature. You are at a meeting where promotions are being discussed and want to hear what they are saying about you. So you find an excuse to leave the room for a few minutes, hustle down to the nearest phone booth and call your phone, which you left lying on a chair, it auto-answers, and you can hear everything being said in the room. And you are being overlooked again. Maybe next year . . .

ROLL YOUR OWN

You spent several hours dickering with Weasel Willie but were not able to get what you need. Nothing that you have read about so far appeals to you. You have some knowledge of electronics and decide that you want to build your own bug. So where do you start?

There are transmitters in kit form that you can build. Technically, a kit can be anything from the old VT-75, which required only that the user solder a few wires to the ceramic base, to a bag of parts and a diagram. Or maybe just a diagram. No parts, no manual.

The Electronic Rainbow in Indiana sells complete kits—just place the components in the board and solder them in place. They are low cost, and they work as well as some others that cost three times as much. Several are listed in Appendix C: Places To Get Things.

Some kits are "from scratch." You have to make your own circuit board, which consists of placing an adhesive film over parts of a copper-plated circuit board and then etching it with a mild acid. It isn't all that difficult to do, but a little practice is needed to get the film in the right places.

You really want total security? Try having your next business meeting aboard a boat in the middle of a large lake, and make sure the vessel is equipped with Sonar that detects SCUBA divers and remote-controlled submarines.

An "Underground" Book from Paladin

For those of you who are interested in the technical details of transmitters, there's a book available from Paladin Press called *The Basement Bugger's Bible* by Shifty Bugman. I had the pleasure of proofreading the manuscript and found it to be very well written and organized, with a wealth of information on all aspects of building transmitters. It is written on the technician-engineering level and is a must-read for anyone serious about designing and rolling their own.

Mr. Wilson

John S. Wilson Jr. has a number of plans for transmitters—dozens of them. I was sent review copies of five of these plans, and all of them were simple modulated oscillators using a single PNP transistor in a voltage divider configuration. These are plans only, but they do have a parts list with the part numbers you need to make sure you get the right ones, as well as assembly instructions. You will need to etch the circuit board and drill the holes, along with mounting the components, to build them. Experience is required. One nice thing about these transmitters is that any of them can be built at very low cost. All the parts and the materials needed to etch the circuit board cost maybe $10 or $15. True disposable bugs. The last address I have is:

Technology 2008
P.O. Box 5264
Augusta, GA 30906

Mr. Arrington

Winston Arrington recently released a revised version of his book *Now Hear This*. It contains dozens of schematic diagrams for room audio and phone line transmitters, dropout relays (that automatically turn on a tape recorder when the phone it is connected to places or receives a call), subcarrier transmitters, a high-impedance phone tap detector, and other goodies.

The transmitters are not cheapo single-transistor modulated oscillators; they are multi-stage devices, some of which have considerable power output and have antenna matching adjustments and audio preamplifier stages. Some very good stuff here. All diagrams have parts values and some have brief explanations, but most have none at all. No circuit board layout is provided. These projects are not for beginners.

The ARRL

There are a number of books published by the Amateur Radio Relay League, which is devoted to ham radio operators. Many are available at the public library, including the *ARRL Handbook*, which has a great deal of info on basic electronic construction, component identification, parts layout, printed circuit board construction, and much more. They also have books on spread spectrum, antennas, and transmitters/receivers in general. They are not specifically about surveillance equipment, but remember: the principles are the same, and there is a wealth of information in them. The prices are reasonable, too.

At this point you may have found what you believe will do the job. Weasel Willie made you a deal you are considering. A first-year electronics student has offered to build you a few bugs. You are convinced that you can roll your own. So, you think you are ready to bug the mayor's office? You are not. You still have a few things to learn. Many things to consider. Let us look at some of them.

DETERMINING THE FREQUENCY

In many situations there is little choice; you may not have a large selection of transmitters and will have to use what is available. There may be situations where this is sufficient to get the job done and others where you will need to obtain a transmitter for different frequencies. Or, perhaps you can improvise and use something from the previous chapter. In making this decision, please consider the following.

Subspace Interference and Air Pollution

There are things that can interfere with reception at the listening post. If you are very close to the target area, then reception will usually be good enough that you probably don't need to be concerned. If the transmitter is a long distance away, and the signal is weak, you might. The worst offender is probably the computer. As you can read in *Digital Privacy* (Paladin Press), computers generate signals all over the spectrum, from a few KHz to 500 MHz. Having a computer operating in the same room as the surveillance receiver is not a great idea. Try turning it off and see if reception improves. As well as disrupting your brain wave patterns, a TV set can interfere with reception, so if you have one at the listening post, turn it off when you are monitoring. Better yet, kick the TV habit and donate it to a homeless shelter.

Electronic telephones and even digital clocks can interfere with reception. My Panasonic Auto-Logic phone/answering machine/etc. causes all kinds of interference to the PRO-2006 scanner on some frequencies. If something is making weird sounds on the surveillance transmitter frequency, try turning things off and see if it goes away.

One particular surveillance trick that makes a transmitter more difficult to find is to set the frequency very close to a strong signal in the target area. The sound portion of a TV signal, for example. This is called "snuggling" and makes it difficult to locate without anything less than a good countermeasures receiver or spectrum analyzer.

Outside interference is an even bigger problem, as it is much stronger, and there is little you can do about it. The worst offenders here are often paging transmitters. They sometimes have a high output power, and there are so many of them that the airwaves are jammed with their transmissions. These signals may consist of a variety of different tones, chirps, beeps, buzzing sounds, and whatever else. There are also zillions of voice pagers that repeat a short message over and over. Dr. Greenbuckle, please call admitting . . . Dr. Greenbuckle, please call admitting . . . Dr. Greenbuckle . . . The problem is not the direct signals; it is something called "intermod." The pager transmissions "mix" with other signals, producing "sum and difference" frequencies, and there are so many of them that they may appear at any place in the spectrum at different times. Totally unpredictable. Could drive a spy into getting an honest job.

Intermod Defined

Transmitter A is operating on 150 MHz.
Transmitter B is on 200 MHz.
When both are transmitting at the same time, and they are within a certain geographical distance of each other, these two signals combine to form two new frequencies known as "sum" and "difference". So we have

150
200
150 + 200 = 350
200 - 150 = 050

Next, all of these new frequencies mix together to create many more new ones. For example, 350 MHz + 200 MHz = 550 MHz, and so on, ad infinitum. Fortunately, most of them are very weak and don't radiate very far, but if you happen to be close to them, they may be a problem. If you should happen to have, or are able to purchase the Xplorer from Optoelectronics (expensive at $900, but an excellent investment), use it to field test. Wander around between the target and possible listening post locations at various times. See what you hear.

If you buy a variable frequency transmitter, read whatever instructions that come with it, if any. They should explain how to set the frequency. This adjustment will be a tiny little round thing that has a slot in it for a screwdriver-like tuning tool. It may also have a second such component which will probably be an antenna matching circuit. FM broadcast band types are usually preset near one end of the band, usually the lower. Fire it up, set an FM radio at the low end, around 90 MHz, turn the volume up, place the transmitter near it, and slowly turn the frequency adjustment screw until you hear a squealing sound called feedback.

If possible, use a nonmetal tool to make the adjustment. Metal, particularly ferrous (containing iron) metal such as steel, can have a slight detuning effect, which may require that you touch it up several times. Plastic tuning tools are available at most electronics and ham radio supply stores for a buck or two. Or, you can improvise—make it from one of those chopsticks you never learned to use very well. They're in the right-hand drawer behind the silverware tray.

For transmitters that operate outside the FM band, the VT-75, for example, you need to modify the radio or use a scanner or communications receiver that receives wideband FM. The procedure is the same. Once you have found the frequency that the transmitter was set on at the factory, you can adjust it to where you want it as shown in the section "Using a Scanner," below.

To change the frequency of a crystal-controlled type, you have to install a different crystal. This requires some experience. Not a good idea to try it if you don't know what you are doing. Then you should know that crystals are not made to oscillate on the actual transmitting frequency; this frequency is increased by one or more circuits called "doublers" and "triplers," which double and triple the frequency. A transmitter that operates on 120 MHz might have a crystal cut for 10 MHz that is doubled to 20 MHz, then again to 40 MHz, then tripled to 120 MHz. Other crystals are called overtone, which means that they resonate at a multiple of the frequency they are cut for.

In order to replace the crystal in a given transmitter, you have to know which type to get. Another consideration is that you have to order them, which costs money and takes time.

Whatever transmitter, and whatever frequency you have decided to use, do some testing over a period of time (if you have the time), and see if anything is interfering with reception.

"Oh, Max, you fell for the old bug in the 'martini olive' trick."
According to *Time* magazine, this actually was used by a Russian agent back in the sixties. It is not recommended if the spy has already consumed several of the potent concoctions.
"Max, the National Battery Ingestion hotline number is 202-625-3333."
"I knew that, 99."

Using a Scanner

Most FM broadcast band surveillance transmitters, wireless microphones, are wideband. They are intended to be heard on an ordinary FM radio. Fine, except that such radios lack the sensitivity of a scanner or communications receiver. They don't have to be as sensitive (which would increase the cost), as commercial FM stations have very high-power signals. If you have a scanner that has wide-FM mode, use it first. Find a quiet spot on the FM radio dial, and make a note of the frequency. Set the scanner on that frequency and adjust the transmitter till you hear it. At very close range, the transmitter can usually be heard on a number of different frequencies (harmonics and subharmonics) in addition to the fundamental one. Try adjusting and retuning it a few times for practice and, if possible, near the target area. Otherwise, you may later discover that it is transmitting on the edge of a 50,000-watt rock station. This may make it a bit difficult to receive.

If the transmitter is VHF or UHF, obviously you need to know the frequency, if crystal-controlled, or the range (window), if it is a variable type. Again, find out before you buy it; otherwise you may have to spend several hours with the scanner trying to find it. Having a frequency counter or test receiver such as the Optoelectronics Scout in such a situation is very useful.

2
How Far Will It Transmit?

This is usually the first thing people want to know when shopping for a surveillance transmitter. As you have read, it is probably not what it is advertised to be. What really matters is not the distance the transmitter can allegedly send the signal to but the distance at which the signal can be heard and understood—something I will call the maximum effective range. This depends on

- the power output of the transmitter
- the frequency it is transmitting on
- the sensitivity of the receiver
- the antennas, both transmitter and receiver
- what is in between the two antennas

Under perfect conditions—line of sight, a good receiver, good antennas on both the transmitter and receiver—and with enough power, a range of several miles is not at all unrealistic for some transmitters. I sent out a preview copy of this book to a number of people in The Biz for proofreading and criticism, and while most had positive feedback, I caught a lot of flack about this statement. One person called me to raise hell, and I tried to point out that I had qualified this statement with "perfect conditions," but all he saw was "miles." Agreed, such conditions are seldom to be found. However, the point I am trying to make is that surveillance transmitters

- do not have to be limited to a range of a few hundred feet
- do not have to be any particular size and shape
- do not have to transmit on VHF or UHF
- do not have to be FM
- do not have to be used with the short wire antenna that they are sold with

And they do not have to conform to a set of standards that most people in The Biz are conditioned to.

31

In using a transmitter, the objective is successful audio surveillance, and there are many different ways to accomplish it. The willingness and the ability to think and act independently, to flaunt convention, and to do things that "just aren't done" can make the difference between success and failure in electronic surveillance. The people who make it to the top in The Biz, the experts, are those who get the job done. Sometimes this requires that they are unconventional. They have imagination. They improvise: they use what is available to them and make the best of it. If you intend to defend yourself against surveillance devices, remember that these are the people you will be up against. The experts don't play by the rules. And if you do, you lose.

Theoretically, a radio signal will travel through space forever, unless something absorbs it. To illustrate this, here is a old-time radio tale I heard a few years ago on a listener-supported FM station.

A man in Texas heard a classical music program that he liked very much, so he tried to call the station to comment. But there was no listing in the phone book, and the operator had no number for the station. Now he knew what he had heard and had written down the call letters, so he was very puzzled about this. But unlike most people who would have just forgotten about it, he decided to check it out. After learning that the station had been off the air for several years, he started making phone calls and writing letters. Finally, he reached someone who had been an engineer at the station. Some time later the two men got together to discuss the incident, and the listener played a recording that he had made of part of the broadcast. Again, this is a story I heard, which may or may not be true. The point is that, in theory, the signal from that radio station, long ago off the air, could have been reflected from some distant object in space, and returned to Earth.

MAXIMUM EFFECTIVE RANGE

We have all heard weak AM stations on the radio, sometimes wishing we could receive them better. They fade in and out (QSB) and are sometimes covered by static (QRM) and interference from other sources (QRN). Receive them better? More clearly? Yes, but we are, in fact, receiving them. Maximum effective range isn't an official technical term; it is just something I use to illustrate a point. If someone tells you that a given transmitter can send a signal 10 miles through space, they aren't necessarily lying to you. But can that signal be received and understood 10 miles away?

Power

Range depends on the power of a surveillance transmitter, which depends on a number of other things. In theory it is unlimited, but in actual use, the physical size of the bug and, often more important, the size of the batteries limit this power. More than one tiny bug has been discovered because the large batteries were so easy to find. Most RF transmitters, the types sold in "spy" shops, will have a power rating of 5 to 40 mw. A comparison: the Spectrum body wire described above is about 250 MW. The homemade AM transmitter used to get the goods on the gold digger in Michigan (described in *Don't Bug Me*) had a power output of about six watts. A portable hand-held radio like police and ham radio operators use will usually have an RF output of 2 to 5 watts and work through a repeater.

Repeaters and Blimp Hangars

If you use a two-way radio on your job, you have no doubt had times when you couldn't get through to raise the dispatcher and other times where you could from many miles away. One job I had a few years back was in an area where there are a lot of mountains, and our repeater equipment and antenna was on one of them. I could access it with my 4-watt portable from about 20 miles away. On the low power setting, 1 watt, I could still hold the repeater from about five miles away. One watt, five miles. But under ideal conditions; the two antennas could "see" each other; nothing was blocking the signal path. At the other extreme, inside a metal frame building, in an office full of metal desks and filing cabinets, my portable could barely reach the repeater.

Under such conditions, the range of a 50 MW transmitter might be only 200 feet or less, and the low-power types of 5 to 20 MW probably wouldn't get out of the building at all. One of the super-cheapos I tested was so poor that if it were on top of the Space Needle in Seattle, the signal literally would not reach the ground!

Here is another example. A few years ago, some friends and I went to the last (unfortunately) air show at

Moffett Field Naval Air Station. While I was out wandering around looking at (drooling over) all the aircraft, my friends stayed back in the parking area. We were communicating simplex, direct and not through a repeater, with hand-held radios on the two-meter ham band. Reception was fine. Then I moved behind the enormous metal blimp hangar (you have to see it to believe it) and, even though we were only a few hundred feet apart, reception dropped to barely readable. My Kenwood radio put out 100 times as much power as some of the "spy shop" transmitters, and if it won't transmit through a steel structure like the blimp hangar, what do you think these little gems will do in the same situation? But Weasel Willie might not tell you about such conditions when you shell out several hundred bucks for one of his transmitters.

Frequency

Now consider frequency to be used. The higher it is the more the signal is line of sight. Sort of. People may tell you that UHF (300 to 3,000 MHz) signals are strictly line of sight, which in a sense they are, but the implication is that they won't travel farther than you can see. Not exactly. The transponders in some satellites transmit in the microwave bands and have no more power than some mobile two-way radios, yet they can be received from hundreds of miles out in space. There are portable phone systems that use a small (four-foot) dish antenna and work through satellites such as Inmarsat. Line of sight? UHF surveillance transmitters will work over long distances if the antennas can "see" each other.

On lower frequencies, such as the 40-meter ham radio band (7 MHz), ham radio operators have been able to communicate with each other over distances of hundreds of miles using less than one watt. Under ideal conditions, that is—meaning they had directional antennas that were on top of a tower and were carefully matched to the transmitters. Citizens Band (CB) operators using 5 watts of power and whip antennas mounted on a truck are sometimes able to converse over long distances on the 11-meter band (27 MHz). The difference here is something called ground wave. At lower frequencies, radio waves follow the curve of the surface of the Earth. "Ground" wave. This is why it is possible to hear international short-wave broadcasting stations halfway across the world. There is no fine line between where one ends and the other begins. While I was visiting some friends in the Pacific Northwest I could sometimes hear, on their Bearcat scanner, police and fire departments on the 30 to 50 MHz bands from as far away as Ohio. But at 88 MHz, the beginning of the FM band, the distance falls off sharply. What is the most distant commercial FM station you have ever heard?

Range: power, frequency, antennas.

An exercise: Stop reading, if you will, and look around. Wherever you are, find a few places where a surveillance transmitter might be concealed. Think about how you might make the drop in a variety of situations. Learn to think like a spy, whether you intend to bug someone in your own self-defense or are searching for transmitters. Visualize and condition yourself until you automatically scan a room when you walk into it—till it becomes second nature. It takes a while to develop this skill. Unfortunately, as I have discovered now that I am out of The Biz, it also takes a while unlearn it.

3
Antennas in General

Antennas work more efficiently when cut to the right length, when they are the same as the frequency wavelength. This is where it resonates. Above, I mentioned that 7 MHz is 40 m; the wavelength is 40 m. Wavelength is calculated as follows: Divide the frequency the transmitter is adjusted to in megacycles (MHz) into 300. This will give you the length (in meters) of a full-wave antenna. Full-wave antennas aren't always practical, depending on the use. Forty meters is 131 feet, so antennas are usually half wave or quarter wave. Ham operators can often use them stretched between trees or towers, but spies usually do not have such luxuries. So they use quarter or eighth wave.

If a transmitter is set on 100 MHz, divide 300 by 100 and you get 3. This is 3 m, which equals 118 inches (9.83 feet). It might be possible to conceal an antenna this long, but if not, cut it to 59 inches (half wave) or 30 inches (quarter wave).

A long wire, strung around a room, such as under the edges of the carpeting, will sometimes work better than the short length of wire that was supplied with the transmitter. The lower the frequency, the more this is true. Curtain rods make great antennas if they are matched; they have the advantage of already being in place and being near a window, and they are more exposed. Nothing is there to block the signal.

The aluminum siding on a house can sometimes be used as an antenna. While it is supposed to be grounded, it may not be and so will sometimes work, and remember that it is on the outside of the house, whereas a wire antenna is on the inside. In a commercial office building, the heating and air conditioning ducts will sometimes produce excellent results, and other times will not. Experiment if possible. If you don't have the time to experiment (which usually means you are installing the transmitter in a place you are not supposed to be installing the transmitter) and none of the other options are available, then use a wire antenna, but have it precut to the right length.

In the field, you work with what you have in the amount of time you have.

The Antenna Farm

Directional antennas such as the beam (Yagi, and log periodic) concentrate a transmitted signal in one direction. This increases the signal strength as if it were actually a higher power, known as Effective Radiated Power or ERP.

The dipole transmits broadside in two directions and the ground plane or Marconi, transmits equally in all directions. But no matter what type is used, both transmitting and receiving antennas should both be polarized the same way, and if possible matched to the transmitter.

If using a beam, a TV antenna for example, be sure to connect the transmitter antenna wire to the driven element or the signal will be grounded.

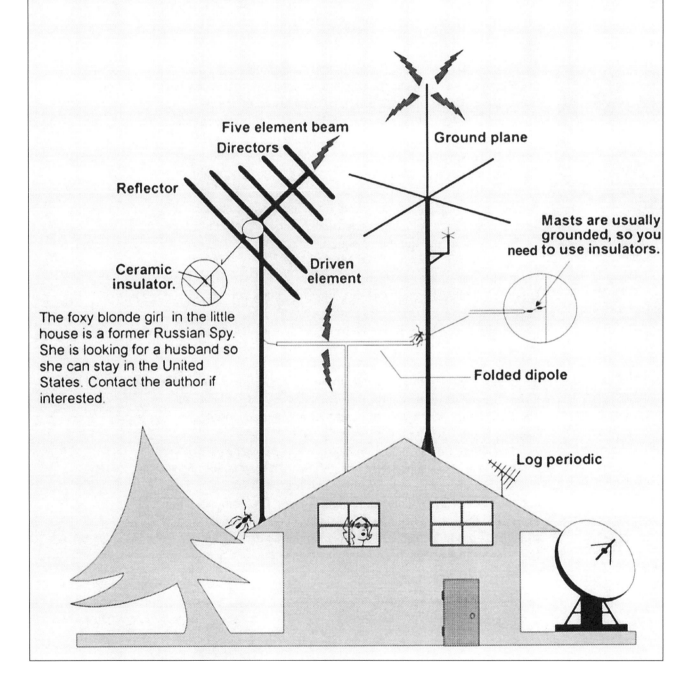

Five element beam
Directors
Reflector
Ground plane
Ceramic insulator.
Driven element
Masts are usually grounded, so you need to use insulators.
Folded dipole

The foxy blonde girl in the little house is a former Russian Spy. She is looking for a husband so she can stay in the United States. Contact the author if interested.

Log periodic

RESONANCE

It is not always possible to use an antenna that is cut to the right length, to its resonant frequency.

But it might be possible to adjust the frequency to the antenna that is available. A grid dip meter, or "grid dipper," is an electronic device used to measure resonant frequency. Intended for use with resonant "tuned" circuits, it works, with varying degrees of success, with various metal objects. Such as the curtain rod. The grid dipper was once found in any ham radio station and radio repair facility, but in recent years they have seemed to disappear from the market. Millen and Heath models may be found in surplus electronics stores, but I don't know who is still making them. If you have the skills, you can build your own. For details, see:

http://hem2.passagen.se/sm0vpo/use/gdo.htm

This idea is "iffy" and controversial, and not usually practical unless you have a lot of time in the target area, but it's something to be aware of.

TYPES OF ANTENNAS

There are many types of antennas; a few of them are

- Marconi
- rhombic
- log periodic
- yagi
- dipole
- cubical quad
- spiral log conical
- J-pole

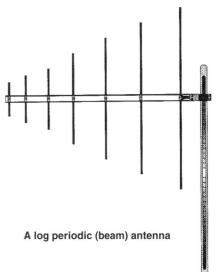

A log periodic (beam) antenna

Most of these are rather uncommon in surveillance, and while they may work fine, there are four types that are used most often.

The Ground Plane

The ground plane (Marconi) is a vertically polarized quarter-wave type used mostly for CB and business band radio and is available at most CB radio stores. If a transmitter is being installed where an unused ground plane is available and you can get to it, then use it. Even though it is cut for a frequency other than what you are transmitting on, the high location makes up for this. The ground plane transmits and receives in all directions. However, test it first—the transmission line may be open or shorted.

The Dipole or Folded Dipole

Dipoles are usually half-wave and are used mostly for amateur radio equipment and FM tuners and receivers. The dipole is horizontally polarized, and it transmits and receives in two directions.

The Longwire

A longwire antenna is just that—a long piece of wire strung in the highest location available. Longwire antennas are used most often for shortwave listening (SWL) in the high frequency (HF) bands but will pick up the signal from an 87 MHz surveillance transmitter better than most inside antennas. The longwire is horizontally polarized and not usually used for transmitting, but with a surveillance transmitter, it can get greater range than a short piece of wire. Longwire antennas are often easy to disguise. Or hide. They can be made of virtually any type of wire, such as telephone cable, speaker wire, or the wires that connect doorbells, thermostats, intercoms, or whatever else.

If you are considering such a wire, first make sure it is not being used for any other purpose. If you see smoke and sparks coming from your transmitter, you'll know that it was.

Beams

Beams are directional; they concentrate the signal in one general direction, "focusing" it toward the receiver. Beams are made up of different parts called elements (see glossary and drawing) and can be polarized either horizontally or vertically. Using a directional antenna, either for the transmitter or the receiver (or both), can dramatically increase the maximum effective range. This is especially important with DSS transmitters.

ERP

This focusing of a directional antenna can increase the maximum effective range of a transmitter because it increases what is known as the Effective Radiated Power (ERP). You may hear some commercial broadcasting stations use this term when they sign off for the night: "station KBUG with an ERP of 50,000 watts . . . by authority of the FCC. . . ."

An antenna that has three decibels (dB) forward gain would double the power. A 100 MW transmitter using such an antenna would have an ERP of 200 MW. So the actual RF power of KBUG might be closer to 25,000 watts.

SPIES LOVE CABLE TV

Well, I don't mean that spies necessarily love to watch cable TV. Perhaps they don't. What I mean is that spies love the idea of so many people having cable TV. This means that all over America, there are millions of unused TV antennas on people's roofs, just sitting there, waiting to be used with a surveillance transmitter. Like maybe yours?

If there is an unused TV antenna at the target, it may well be the best choice. VHF TV antennas will work very well with wireless microphones that operate on the commercial FM band. TV channel 6 is just below FM, from 82 to 88 MHz. As stated before, this may be your best option, but if possible, test it first. Have someone report on reception from the listening post using a two-way radio. Best option? I should qualify that statement. TV antennas are about as well maintained as taxicabs. The lead-in wire insulation is often old and brittle where it has been exposed to the sun. The wires inside might be broken. The aluminum corrodes and the bolts rust. No one ever gets around to fixing them. It "just isn't done."

TV antennas are directional, so if they are pointed toward the listening post they will work better. If possible, turn it to the proper direction. Take some heavy pliers and a crescent wrench. If, for some reason, you cannot rotate it, maybe you can bend one of the elements (the one the transmitter is connected to) in the proper direction. Be careful, as the aluminum elements break very easily. Take a roll of duct tape to secure it.

The same thing is true of UHF TV antennas used with a UHF transmitter. However, they will also work the other way: a VHF antenna with a UHF transmitter and vice versa. They will not work as well, but being in a more exposed location will help make up for this.

And how, you ask, am I gonna get up on someone's roof without having to explain myself to the owners, nosy neighbors, security guards, or cops? One of the secrets in a business where there aren't any secrets is to always be able to have a good explanation for being where you are, or, better yet, have permission to be there. The possibilities are limited only by imagination. You are a contractor who offers free roof inspections. You are an itinerant TV antenna repair service company. You have spotted a yellow-breasted sapsucker on the roof and will pay the people if they permit you to photograph its nest (having what will pass for a *National Geographic* photographer ID or Audobon Society membership card might work). And, of course, offering them $100 for their trouble might do the job. You need to be able to feel people out. Read their expressions and their body language. Know how to come up with the right story. And how to react to seeing a 12-gauge shotgun pointed at the reproductive regions of your anatomy. Being a spy isn't all technical knowledge . . .

RECEIVING ANTENNAS

The antenna used with the receiver is also important, as much as the transmitting antenna. So, as transmitting antennas work best cut to the right length, so do receiving antennas. Also true is that the more exposed they are, the better they work. I bought a telescoping antenna for one of my portable scanners and made some comparisons. With the longer antenna I got better reception on VHF and low UHF than when using the rubber duck that came with it. In the high UHF band, up around 800 MHz and above, the shorter antenna worked better. Remember, most surveillance transmitters operate between 50 and 460 MHz.

An outside antenna makes a very big difference in reception. The public service agencies I could receive on the Bearcat scanner from some 2,000 miles away were heard using an all-band vertical antenna made for scanners, on top of a 30-foot mast. With the inside antenna, I couldn't hear them at all. A directional antenna, such as a beam, will improve reception, as well as transmission, considerably.

So, while you may be very limited in what you can use as a transmitting antenna, at the listening post you usually have more choices. As with all other phases of surveillance, use the best you possibly can. Now, while the type of antenna to be used is important in surveillance, there are some other things that are equally important.

Polarizing

A few years ago I was at the Spy Factory (yes, I spent a lot of time there) delivering some books when a man came in with a complaint. A transmitter he had bought there didn't have the range it was advertised as having (no kidding!), and he wanted his money back. I asked him if he had cut the antenna to the right length, and if both antennas were polarized the same way. He didn't understand, so I explained it to him. So he said he would go home and try this. The clerk didn't tell him anything about antennas because he didn't know these things, and the documentation that came with the transmitter did not include any such information.

For best results, the antennas on both the receiver and transmitter should have the same polarization. If the transmitting antenna is vertical, then the receiver antenna should be too. Ditto horizontal. This can make a very big difference in the effective range, and in many situations, it will be the determining factor in whether the system works or not. Success or failure can depend on such a little thing.

Horizontal polarization means the antenna is parallel to the ground. Vertical is, of course, perpendicular to the ground. It usually makes little difference which you use with a surveillance transmitter, as long as both antennas are the same. However, as an added measure of security, you can polarize them the opposite of commercial radio services. Where most are vertical, make yours horizontal. This decreases the chances of your bug's being intercepted by "unauthorized" persons and might reduce intermod.

Alignment

As you read above, directional antennas focus or concentrate the signal much like changing the focus of a flashlight; some are more directional than others. The sharper the focus the more directional the signal, the more closely the antenna needs to be aligned with the listening post. Because of this, it is important that it be pointed as directly as possible toward the listening post. Generally, just eyeballing it may be all you can do, but if you should have a partner at the receiving end to relay you a reception report by two-way radio, it is possible to zero it in a little more accurately.

Placement

Whenever possible, place the antenna so that it is in a direct line with the listening post and, if possible, with nothing in between. Walls, for example, will cause signal losses where windows will not. Metal objects in between such as filing cabinets or desks will substantially reduce the signal, perhaps to where it cannot be received at all. An outside antenna is almost always better than an inside one, and the higher the better.

Matching

For a bug, for any transmitter, to transfer most of its power to the antenna, that antenna has to be matched to

An inexpensive digital FSM

the transmitter and the transmission line, the cable that leads from the transmitter to the antenna. If the impedance is not matched, the result will be standing waves.

Standing waves are part of the power from the transmitter that hovers or "stands" on the transmission line and never reaches the antenna, so it isn't transmitted. A few transmitters, such as the VT-75, have an adjustment to match the transmitter to the antenna for maximum power transfer, i.e., a lower standing wave ratio (SWR). If you are using a transmitter that has this feature, set it like this:

Using a test receiver, set the transmitter on the frequency to be used and connect the antenna. Now fine tune the radio for the strongest signal and listen carefully to hear if there is any background sound, such as the hissing sound heard when an FM radio is off station. This means the signal is weak.

If it is perfectly still (known as "full quieting"), the adjustment may be set correctly, but also the transmitter and receiver may be too close to each other. Remove the antenna from the receiver or move it farther from the transmitter until there is some noise. Then s-l-o-w-l-y turn the antenna adjustment screw until the radio is completely quiet. Change the distance as needed until you get the best results. Should you get feedback, turn the volume down or cover the microphone with something. Sometimes you may have to move the receiver a few times to get it set to the best position.

The Field Strength Meter

A better way is to use a field strength meter (FSM). Place it a few feet away and adjust the antenna control for the highest reading on the meter scale. If the needle doesn't move (meaning that it is not receiving the signal) place the meter closer to the transmitter. If the needle stays all the way over to the right (signal too strong), move it a little farther away. If you are using an outside antenna, have someone hold the FSM as close to it as is necessary (if possible) to get a reading and report the results to you on a two-way radio.

An FSM can be put together for less than five bucks. The meter should have a range of 0 to 10 microamperes (µa) DC, and most RF diodes will work. Anything made of metal can be used as an antenna, but the telescoping type, removed from an old radio, is suggested because you can extend and collapse it. The closer to the transmitter you get, the higher the reading on the meter will be. Used carefully, it will find transmitters as well as most "bug detectors" in the $50 to $100 range. A decent FSM, in kit form, is available from Alltronics, listed in Appendix C.

4
Electronics 105—Ohm's Law

For your disinterest, here is some stuff on basic electronics that will help you in understanding battery life and transmitter power. If it doesn't interest you, you may be tempted to skip it. And then later, you'll wish you had read it.

VOLTS, AMPS, OHMS, AND WATTS

Ohm's law is the most basic formula in DC electronics. It is the relationship between current, voltage, and resistance and is expressed as

R = E divided by I
I = E divided by R
E = I multiplied by R

I is electrical current, and the basic unit is the ampere. The ampere (commonly called the amp) is also expressed as the milliampere (ma), which is .001 amp, and the microampere (ua) which is .001 ma or .000001 amp.

E is voltage, the basic unit being the volt. The volt is also expressed as the millivolt (mv), which is .001 volt, and the microvolt (uv), which is .001 mv or .000001 volt.

R is resistance, and the basic unit is the ohm. The ohm is also expressed as the K ohm, which is 1,000 ohms, and the megohm (MΩ), which is 1,000 K ohms or 1 million ohms.

Current (I) is the movement of electrons through a wire. Voltage (E) is the pressure that makes the electrons move. Resistance (R) restricts the movement of electrons.

The flow of current through a wire is analogous to water flowing through a pipe. Voltage is like gravity (from a water tower), which causes the water (current) to flow. Current is the movement of water (electrons) through the pipe (wire). Resistance is a valve in the pipe (wire) that controls how much water (current) flows, as does the size or capacity of the pipe (wire). A big pipe can move more water than a small one, and the same is true of a wire. If the pressure (voltage) in a pipe is too great, it will burst. If the current in a wire is too great, it will get hot and eventually melt. If the current is very high, as in a direct short in the power line, the wire will vaporize instantly. It can also vaporize your fingers, that is, cause some very serious burns, so don't mess with 110 volts from the power line unless you really know what you are doing.

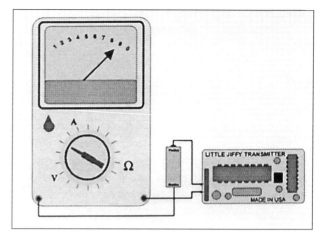

From the formula E = I*R, you can see that if a 1-volt battery is connected to a 1-ohm resistor, then 1 amp of current will flow. If you double the voltage, the current doubles. Double the resistance and the current is cut in half.

Power is the product of current (I) times voltage (E).

In this example, the voltage is 1 volt, and the current is 1 amp; 1 volt times 1 amp is 1 watt. One watt of power is being consumed from the battery. Battery power is rated in ampere hours; that is, how many hours a battery can produce a current of one ampere, or a fraction thereof.

Efficiency is the ratio of power out versus power in. No electrical device is 100-percent efficient. A motor converts electrical power to mechanical force, but some of this power turns into heat. Motors get hot. A light bulb is intended to produce light, but much of the power also produces heat. Light bulbs get very hot and are very inefficient. This is also true of surveillance transmitters. A transmitter that consumes 100 MW of power and has an output of 50 MW is 50-percent efficient. Typically, they are not quite that efficient at low power levels.

To measure the current a transmitter is drawing, connect the meter test leads as shown in the diagram. Set the meter on the DC current scale on the range of 0–30 ma, or whichever range is closest to that. Write down the readings every hour or so. Switch the meter to the DC volts scale and measure the battery voltage. Check it now and then, keeping track of the elapsed time. When the voltage reaches the cutoff point (next chapter), the elapsed time is about what you can count on when using the transmitter in the field.

5
The Secret Lives of Batteries

If you want to know about bugs and how to use them, you need to know about battery life. Assuming that you will be installing a battery-powered transmitter, it is essential that you know how long the batteries will last.

You have been hired to bug the boardroom of Wexler's Widget Works to eavesdrop on an upcoming meeting. You have the date and time, but your access to the room is limited so you need a transmitter that will still be operating during the meeting. Otherwise you fail, and instead of being paid a lot of money, you will receive a pair of cement shoes. And an invitation to go swimming in the Hudson River that you cannot refuse.

Battery life depends on how much current the transmitter draws (power consumption) and the type of battery used (power available). Most surveillance transmitters don't even mention this in the documentation. Suppose some marketer claimed one of his transmitters has a range of half a mile and a battery life of about three weeks from two AA cells. Let us assume this means that the signal can be received and understood at that distance. Maximum effective range.

I will use this as an example of how to understand and estimate battery life for a particular transmitter. Let's make some calculations. They will be based on spec sheets for Mallory batteries, courtesy of Energy Sales Co. of Mountain View, CA, and Eveready batteries courtesy of the Eveready Battery Co. in south San Francisco. The previous chapter, on Ohm's law, will help to make this easier to understand. (I asked you to please read it, but you wouldn't listen, would you?)

An alkaline AA cell produces 1.5 volts, with a rating of 2.45 ampere hours. This means it can theoretically produce a current of 2.45 amps for one hour, 245 ma for 10 hours, 24.5 ma for 100 hours, or 2.45 ma for 1,000 hours. In reality, it is a little more complicated than that. The actual rating is based on a load (the device that it is connected to) of 24 ohms and a drain current of about 50 ma. If you draw 2.45 amps, the battery may reach the cutoff in less than one hour, and if you draw 24.5 ma, it may last longer than 100 hours.

It also depends on other factors, such as temperature and duty cycle. Duty cycle means how long the device is operating compared to how long it is turned off. Used continuously, the life is shorter than if it is used intermittently. If it is allowed to rest periodically, it will regenerate to some extent. You may have experienced this with a flashlight. It starts to get dim, but if you turn it off for a while, when you turn it back on it is a little brighter. It is the same with transmitters. If a device, whatever type, is turned on all the time, it has a duty cycle of 100 percent.

Battery life is based on the rated cutoff voltage, which is the point where the battery is no longer able to operate the device it is powering. If you have a portable radio that uses AA cells, when these batteries discharge to about 0.7 volts, the radio will lose most of its audio volume, if it even works at all. At this voltage, flashlights become very dim, and electronic flash units, like the one used to take some of the photographs used in this book, usually quit working.

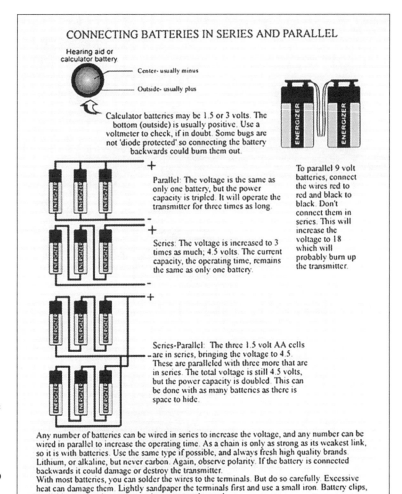

CONNECTING BATTERIES IN SERIES AND PARALLEL

Hearing aid or calculator battery

Center- usually minus

Outside- usually plus

Calculator batteries may be 1.5 or 3 volts. The bottom (outside) is usually positive. Use a voltmeter to check, if in doubt. Some bugs are not 'diode protected' so connecting the battery backwards could burn them out.

Parallel: The voltage is the same as only one battery, but the power capacity is tripled. It will operate the transmitter for three times as long.

Series: The voltage is increased to 3 times as much; 4.5 volts. The current capacity, the operating time, remains the same as only one battery.

Series-Parallel: The three 1.5 volt AA cells are in series, bringing the voltage to 4.5. These are paralleled with three more that are in series. The total voltage is still 4.5 volts, but the power capacity is doubled. This can be done with as many batteries as there is space to hide.

To parallel 9 volt batteries, connect the wires red to red and black to black. Don't connect them in series. This will increase the voltage to 18 which will probably burn up the transmitter.

Any number of batteries can be wired in series to increase the voltage, and any number can be wired in parallel to increase the operating time. As a chain is only as strong as its weakest link, so it is with batteries. Use the same type if possible, and always fresh high quality brands. Lithium, or alkaline, but never carbon. Again, observe polarity. If the battery is connected backwards it could damage or destroy the transmitter.

With most batteries, you can solder the wires to the terminals. But do so carefully. Excessive heat can damage them. Lightly sandpaper the terminals first and use a small iron. Battery clips, available at most electronic supply stores, can be used.

SPY TRICK

Always use a fresh alkaline or, better yet, lithium battery and stick with brand names. Eveready and Mallory are good. Other brands might be as good or even better, and they might not. Radio Shack carbon batteries should absolutely never be used. Why take the chance just to save a few bucks?

For this example, let us base the calculations on the numbers from the spec sheet: 50 ma into 24 ohms. At a current drain of 50 ma, the battery will reach the cutoff voltage (0.8 volts) in about 50 hours. The power output of a bug that is using 1.5 volts at 50 ma is approximately 37.5 mw (50-percent efficiency). A half-mile range at this power level is possible, under very good conditions.

Now, three weeks, as advertised, is 504 hours—10 times as long as the AA cell can produce 50 ma. This means that if the battery is to function for 504 hours, it can produce only about one-tenth that amount of current, or about 5 ma. The power output of a bug that is using 1.5 volts at 5.0 ma is approximately 3.75 mw (50-percent efficiency). A half mile at 3.75 mw? Not likely, even under perfect conditions. OK, now this transmitter uses two AA cells. The ad doesn't say whether they are used in series or parallel. In parallel, the power output will be the same and the operating life will be doubled. In series, the power is doubled and the battery life is the same. To give this transmitter the benefit of the doubt, I will assume they are used in series. The power will be doubled to 7.5 mw. Half a mile at 7.5 mw? I seriously doubt it. The docs that come with the

VT-75 claim a range of 1/3 mile with 75 mw—10 times as much power. However, according to electronic theory, this range is possible. Based on a receiver with 1 microvolt sensitivity, with perfectly matched antennas, allowing for free space losses in signal strength, loss from obstructions such as walls, trees, and so on, 12 mw can transmit a readable signal half a mile. But in real life it doesn't often work out that way.

An engineer friend of mine works for a company that measures radiation from various types of electronic equipment to make sure it complies with FCC regulations. We worked out some calculations, allowing for these losses, and then he went out and made some tests using sophisticated absolute field strength measuring equipment. Reception was far less than the calculations said it should be. The reason for this is very complicated, and I will not go into it. I don't really understand it. Suffice it to say that the only way to be sure of the range is to field-test it.

The target is the reception area of an insurance company. You may get one or more chances, depending on how you handle the situation. You might go in pretending to want to buy insurance, or to apply for a job. The transmitter to be used is in your pocket with the antenna wire taped to your leg. At the listening post an associate is waiting to see if the signal comes in. He then calls your digital pager and leaves a prearranged code number to tell you how well it is being received. Field-test whenever possible.

BATTERIES GETTING WIRED

- Transmitting time can be increased by connecting a number of batteries in parallel. As long as there is a place to hide them, there is no limit to how many can be used. Also, it makes no difference what size is used, as long as the total voltage is what it is supposed to be. Several C cells could be wired to a few AA cells, for example.
- The amount of current a transmitter draws is determined by the supply voltage and its internal resistance. Self-regulated. So using a large type, such as a car battery, will not damage it.
- The power output of a transmitter can sometimes be increased by connecting batteries in series. If a transmitter operates from one 1.5-volt battery (AA, C, D) it is probably safe to increase this to 3 volts without damaging it. Increasing it to 4.5 volts will likely burn it out.
- If it operates from a 9-volt battery, it is probably safe to increase the voltage to 12 (or 13.8, which is the actual voltage of a car battery), but two 9-volt batteries in series (18 volts) will almost surely burn it out. The VT-75, for example, runs very hot at anything over 13.8, and sooner or later it will start to smoke. I burned up a few while field-testing them. There is no way to know for sure without trying it, but before you take the chance of destroying the transmitter, why not try the other things that can improve performance, such as a better antenna, a better receiver, etc.?

THE UN-BATTERY

A great way for an amateur spy to ruin his own day: sneak into the target area . . . install the transmitter . . . get the hell out of the place . . . hurry back to the listening post . . . turn the receiver on . . . and hear . . . buzzzzzzz.

Power supplies—those plug-in adapters that convert 110 volts AC from the power lines to 6 or 9 volts for tape recorders and radios—can also be used to power surveillance transmitters. Some work well, but many of them will not. The reason is that these adapters use very small filter capacitors, so the AC current isn't completely converted to DC. A certain amount of AC component, known as "ripple," gets through, and this causes the buzzing sound. In extreme cases, it makes them virtually useless. This can be corrected by using a larger value capacitor, but as the value increases, so does the physical size. If space is not a problem, then use all the capacitors you need. Install the transmitter inside something that plugs in, such as a clock, lamp, etc. Table lamps are a favorite because there is so much space inside most of them. A well-equipped spy will have several of these adapters and filters in his bag of tricks, configured in different ways. Meanwhile, test it before you install it.

Now these things are so commonplace that some people are used to seeing them without seeing them. It is very possible that if you use one for a transmitter, it might not be noticed for some time.

On the other hand, someone might spot it right away. This depends upon where it is placed, how visible it is, and how observant someone is. Plugged into a wall socket in plain sight, it will be spotted more quickly than if it is in a socket that is behind a sofa, chair, desk, whatever. If "buried" amongst the rat's nest of wires of a complex computer system (you should see the cables under my desk!), it may remain undiscovered for years.

One way to make the device less noticeable is to open it up and remove the brass prongs that go in the socket and solder a line cord in place. An ordinary 100-volt plug is less likely to be spotted. Another variation is to obtain one of those little plastic plugs used to block the openings in the socket. They're used to keep little kids (and dumb adults) from sticking metal objects in the outlet and getting zapped. They are available at hardware stores. Replace the plastic prongs with the brass ones from the adapter and carefully solder the two wires of the line cord in place. Use small wires, making sure they are strong enough that they won't break; conceal them as best you can; and lead them to where there is space for the adapter, filter, and transmitter. Having such a gimmick in your spy kit is a good idea. Remember: you work with what you have, in the time you have.

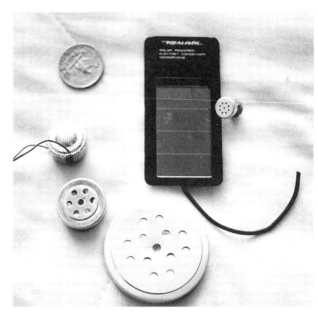

Different types of microphones. The large one on the bottom is crystal. The others are electret. They require a small electric current to work, except the square one, which is solar-powered.

Also, keep in mind that you are working with 110 volts, which can be very dangerous. Not only can you get zapped, but if the wires are shorted, you can get a really nasty burn. This is also a potential fire hazard. If you don't know what you are doing, don't mess with it.

SPY TRICK

Keep a few squares of double-sided adhesive material, such as Scotch Mount, in your kit. With it, a small transmitter can be stuck to a window frame or under a desk in seconds. Remember the "inventor"?

THE SUN BATTERY

Solar cells can be used to power a transmitter if it is in a place where it gets enough sunlight. In some situations it is possible to place the transmitter inside the room to be monitored, and the solar cells outside. Very fine wires can be used to connect them together. It is also possible to place the transmitter and the solar cells outside, and only the microphone inside. The microphone is generally much smaller than the transmitter, so it is more difficult to spot, although it will be found in the physical search. Remember: the transmitter and the microphone do not necessarily have to be in the same location.

6
Room Audio or Phone Line?

Some surveillance transmitters are made for use only with a microphone. This is sometimes called room audio or room surveillance. Others are for the phone line—wiretaps. Let's start with room audio.

ROOM AUDIO

Remember the claims made in Truth in Advertising? "Pick up a whisper at fifty feet." Yet another advertisement I received from a mail order company (I used to get a lot of them) said that their transmitter will pick up sound 100 feet away. It didn't say anything about what kind of sound. Even the cheapest models will hear a jackhammer at 100 feet, but normal conversation at that distance isn't likely. Single-stage (one transistor) modulated oscillators don't usually have very good audio. Better transmitters, with three or more transistors, usually have an extra audio preamplifier stage. The old VT-75 was one such transmitter, and the Xandi model XFM100 is another. Some of the Cony models have nice audio. A Ruby brand (made by Lorraine in England) that I used for a short time, borrowed from the old Spy Factory, had excellent audio sensitivity and quality.

The following diagram shows the International MicroPower WM-1. If you understand schematics, compare it with that of other transmitters. The 14-turn pot makes changing frequencies more precise, and it has an antenna-tuning capacitor for more efficient power transfer.

It has the best audio of any transmitter I have tested. I placed it in a closed drawer, under a pile of towels, and turned a radio on to a normal listening level. Then I went into the computer room and tuned in using a PRO-2006 scanner. The sound quality was outstanding because it has a pre-emphasis circuit. Low frequencies make up more of the "power" of an audio signal than the high

47

frequencies, which cause the sound from a surveillance transmitter to sound muffled, as if "coming from inside a barrel." Pre-emphasis gives more "weight" to the higher portion of the audio range, which makes the received signal easier to understand.

Unfortunately, they went out of business about eight years ago, but there may be a few of them available, and if you look hard enough you might find one. See Weasel Willy.

The point here is that while maximum effective range is important, so is being able to understand the intercepted conversation. There are situations in which even the best of transmitters will not work very well (if at all), such as in an area where the acoustics are very bad and there is a lot of background noise. In such a situation the microphone has to be very near the person to be intercepted. Get as close as possible, and use special audio equipment that may improve the received sound by filtering out some of the noise. An equalizer, of the type used with stereo equipment, and digital signal processing circuits can also "clean up" the sound. Viking International in San Francisco once had some equipment that was useful, but I don't know if they still do.

USING A TRANSMITTER ON A PHONE LINE

Before you install a phone transmitter, you need to know how to connect it. It will be either of two types: series or parallel.

Most of the transmitters sold in "spy" shops and by mail order are series. Either type will have two wires that go to the phone line, which are connected as shown in the drawing. If you have a transmitter and don't know which type it is, try it as a series first and see if it works. Reverse the two wires if necessary. If you test it first on your own phone line, you will already know this and save precious time.

Now, before you install it, you need to decide where you are going to access the line. Follow the wire to what seems to be the safest place, also considering the location of the antenna, if the transmitter uses a separate antenna, and the listening post. The transmitter should already be adjusted to the frequency you are going to use, and you should have a receiver with which to test it. Remember that the idea is to get in and out as fast as possible.

The cable to a single line phone will usually be beige and about a quarter inch in diameter. Inside will be four small wires with red, green, yellow, and black insulation. The wires to be used will normally be the red and green ones. To install a series type, connect the two wires from the bug, about half an inch apart, to one of the phone wires. Try the green one first.

Don't cut the phone line yet. This is to avoid interrupting phone service. If someone was making a call at the time you were installing the tap, they might become suspicious and go have a look.

Would you be able to explain what you were doing there? Do you have a cover story prepared? A maintenance worker's uniform? A clipboard with some official-looking work orders? A passable telco ID card? Someone who does desktop publishing can easily make an official-looking card. Do you know what a real one looks like? Neither do most people. But good spies don't take bad chances.

Once the connections are made, then cut the phone wire between the two connections. Listen for the dial tone coming from the earphone of the test receiver. If it doesn't work, reconnect the phone line where you cut it, and reverse the red and black wires that go to the bug. If this fails, repeat the

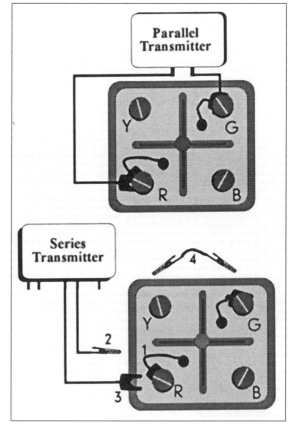

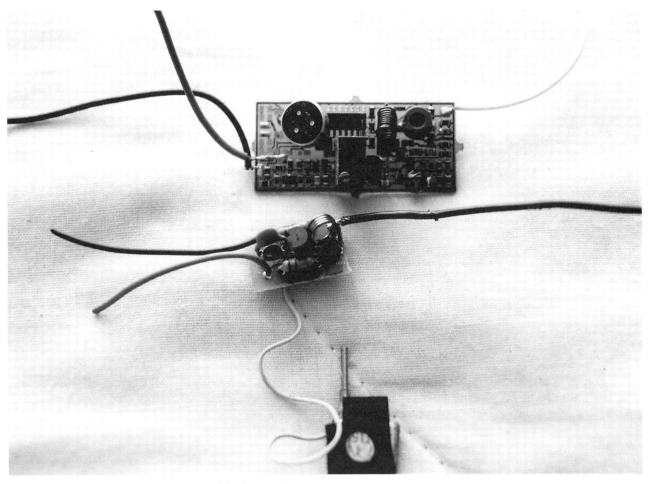

Top: This is the International Micropower FM wireless microphone.
Bottom: A single-transistor modulated oscillator with attached microphone. The sound opening on the microphone is less than 1/8 inch. For serious surveillance, this bug is virtually useless.

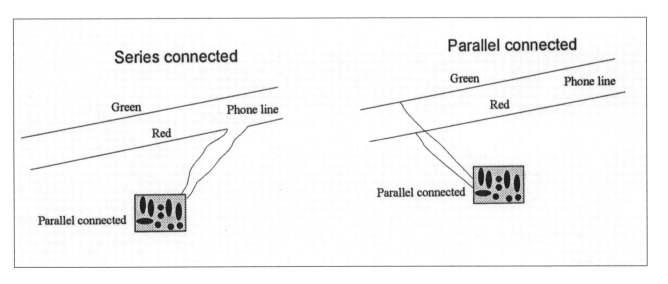

process using the red phone wire, again remembering to first reconnect the green phone wire that you cut. Once you have the right wire and the right polarity, then use Scotchlok fasteners, and conceal the tap as well as is possible. If the wiring you are tapping into is the older type—two large wires with heavy black insulation—found in older homes and apartment buildings, the process is exactly the same.

To install a parallel device, just connect one of the transmitter wires to one phone wire and the other to the other. Usually they, the transmitter wires, will be red and green or red and black. The red one should go to the red phone line, but if it doesn't work, reverse them. One of the transmitters that Sheffield made has an LED to show if the polarity is right.

SPY TRICK

When installing a phone tap, use a single earphone, not headphones, with the test receiver. If someone approaches, you will better be able to hear them coming and get the hell out of Dodge.

OK, so now you know the basics of wiretapping. If you want to learn more, read *The Phone Book*. Meanwhile, don't tell anyone. It makes people nervous. I see the reactions when people ask what kind of books I write. Next, a look at some things you can do with transmitters, other than to bug thy neighbor.

7

Legal
Applications
for Transmitters

Various laws say one cannot use a radio transmitter to bug thy neighbor—and apparently make it unlawful to use or even possess devices made primarily for that purpose. But there are many other uses for transmitters that could be used as bugs, that apparently are not against the law. The big gray area here is intent: what is intended as a bug and what is not. Wireless microphones (i.e., microphones without cords, used by stage entertainers) can also be used as surveillance transmitters. Following are some other examples.

A SECRET RADIO SYSTEM

Some transmitters operate outside the 88 to 108 MHz FM band, so people listening to FM broadcast radios aren't going to hear them. Using two transmitters and two scanners or modified FM broadcast radios, two people could set up a short-distance communications system for under $100. Quite a bit more private than CB or commercial 450 MHz radios, and, of course, some people like to build things themselves.

A BURGLAR ALARM

A transmitter can be set up inside your business or office and be activated at closing time. If you live within its range, you will hear a burglar trying to break in. You can wire a switch to the transmitter so that it will start sending only when that switch is closed. This switch can be mounted on the office door, a safe, or a filing cabinet—wherever you want to place it. It could also be wired into the dome light switch on a car door to prevent someone from ripping off your Blaupunkt and Deep Purple CDs . . .

THE PHONE GUARD

Attached to a phone line, a transmitter will alert you if it is being used by some unauthorized person. A few years ago, I got a call from the Spy Factory. The owner of a construction company was in the store asking about phone transmitters. It seems that someone in the office was using the phone for long distance calls, and he wanted to find out who it was.

I got together with the guy and suggested a VT-75 transmitter modified so that it would trigger an alarm whenever the phone was used. We set it up with a portable FM radio in his pickup truck. When the signaling device came on, he could go inside and see who was on the phone.

It was set up this way to try to avoid breaking surveillance laws; the actual conversation was not intercepted, and it was installed on his own phone line. Apparently he was able to resolve the problem, as I never heard from him again. That's the way it is in The Biz. If the system doesn't work, people come back, but if it does, you don't see them again. The same transmitter could have been set up with a tape recorder that would have turned on whenever the phone was used, and recorded the conversation. But that would have been unlawful if it did not comply with the all-party-consent law.

There are various ways to do this yourself without actually hearing the conversation, and so, presumably, stay within the law. For one, the audio output of the receiver used to monitor the transmitter can be connected through a diode (changes it to DC) to an LED which, if the volume is set high, will illuminate it.

Incidentally, if you want to know when someone is using a phone line while you are within the premises, you can buy a Tap Alert that will light an LED whenever the receiver, on the line to which it is connected, is lifted. But this will not detect most taps.

A PERIMETER ALARM

A transmitter placed outside your home could alert you if a prowler should come within microphone range. While a Cony model MM-18 was being tested and then installed on the back porch of a friend's house, some guy came into the yard and started helping himself to the aluminum cans in the recycle bin. I heard him on the scanner and went out and chased him away.

It could also be connected to the front gate so that you would be alerted whenever someone opened it.

"Clem, where the hell do yew think yore goin'? Yew think yore sneakin' out to the tavern agin? Yew better git yore ass back here right now."

"Yes, Dear. Yes, Sadie."

AN EMERGENCY SIGNALING SYSTEM

Another use would be as an emergency signaling device. In some situations where employees are working alone around dangerous machinery, a transmitter is a very inexpensive and reliable way for them to call for help should they need it.

It can also be used by kids to signal their parents if some creep attempts to molest or abduct them. This sometimes happens very close to home, in which case having such a transmitter could literally make the difference between life and death.

One of the products in Appendix C would work, and an experienced technician could modify it to increase the range.

8

How to Install
a Bug

OK, enough theory, let's look at actually using transmitters. This can be a very complicated business, as many things have to be, or at least should be, considered. And while we have learned that certain principles apply to all surveillance transmitters, each situation is unique. What is most important in one may be less in another, and one thing often depends on another. In this section we will get into the world of hands-on electronic surveillance—the way it really works, using what you have, in the time you have.

THE PERFECT SITUATION

The perfect place to bug might be something like this: The target is the bedroom in an apartment on the top floor of a tall building. It has thick carpeting and heavy drapes and an acoustic tile ceiling. There is an unused TV antenna on the roof, and the lead-in wire terminates in a wall outlet, which is only inches away from a power outlet. You have access to the area for as much time as is needed, have an SWR bridge, and the transmitter has an antenna-matching adjustment. You have a converter to change the 110 volts AC to the DC voltage needed for the transmitter, and finally, you have a filter to smooth out the AC ripple to prevent *humm* and keep the signal from getting into the power line. You can hide the transmitter inside the TV outlet, and there is enough space inside the wall for a compression amplifier which improves the audio sensitivity. You can string the power wires behind the wall from the outlet to the antenna wire box, and adjust the SWR bridge for maximum power transfer.

At the listening post, you will be using an ICOM R-7000 communications receiver with a rotating five-element beam antenna on top of a 200-foot tower. At a distance of a mile or so, reception will be full quieting and the audio superb. As in surveillance TV style.

Now we leave the fantasy and return to real life. The successful installation of a surveillance transmitter is not always easy. The conditions are sometimes very difficult to work under, the very best equipment is not always available, frequently there is very little time (none of which is available for mistakes or hesitation), and finally, there is always the risk of what can happen if you get caught in the act. Like maybe being beat up, blown away, or busted. Getting out after a successful drop is a real rush; it's exhilarating as hell. But glamorous it ain't.

RECON—CASE THE JOINT

It is always a good idea to get into the target area and have a look if you can do so without compromising the operation. See how easy it is to get in, what is there, what a transmitter could be hidden in, if there are hanging plants or pictures on the walls, or upholstered furniture . . .

What are there a lot of, another one of which wouldn't be noticed?

In a warehouse filled with cardboard boxes, one more might not be noticed for days or weeks, depending on when they are to be shipped. Can you get inside information? Could you arrange for a box to be there with instructions not to ship it before a certain date? What kind of fire extinguishers do they have? Could you get in and replace one of them? This can be one of the most effective methods of bugging someone; there is enough space inside for a huge bug, a full watt if necessary, and enough batteries to operate it for weeks or even months.

Periodically, they are required by law to be inspected. So, when the batteries do finally discharge, you don a realistic looking uniform, waltz in and replace it with a new one. Such technicians are like mail deliverers and couriers; no one pays much attention to them as they look like they belong where they are—they have a reason to be there that isn't likely to be questioned or challenged. Are there security guards all over the place? What about TV cameras? Just because you can't see them doesn't mean there aren't any that can see you. A good cover will get you past them in many situations, but don't try this at the Federal Reserve Bank.

Check out the perimeter. Look for windows that face the listening post. Can they be opened from the outside? Do they have heavy drapes? Are there bushes in front of them? Is there an electrical outlet on the outside wall? Is there a telephone connection block (SPSP) on the outside wall?

The more you know about the area, the better prepared you can be when it is time to go in.

Visualize

Sometimes you only get one chance; you don't get to see the area before you go in for the drop. In such a situation, you can sometimes visualize what the target area looks like, what will be there, and so on. In an insurance company office there will be the usual things that are in an insurance company office: filing cabinets, desks, office machines.

Here is an example of thinking ahead, of planning and visualizing. You want to get the goods on someone—medical information. You know that he has an appointment with Dr. Greenbuckle, and you plan to intercept the doctor-patient conversation. One doctor's examining room will have much the same things as another doctor's examining room. A box of cotton balls could conceal a powerful transmitter and several D cells. However, a lot of power would be needed, since these little rooms are usually well inside the building and don't have windows. There is also a lot of metal (cabinets, equipment) that could block the signal.

So, what's a better way? Bug the doctor's office, where there probably are windows. You may not be able to hear the actual examination, but when this is done, the doctor will probably have a consultation with the patient, or dictate notes for the file. In this case, you may well get the information you are after. Visualize. Think. Improvise.

Suppose you want to get some inside information on a new high-rise building that is under construction. There will be an office of some kind set up at the site, probably in a trailer that was towed in. And what will always be in such a place? Blueprints, usually rolled up with a rubber band around them or inside a plastic tube. Would an extra roll of them be noticed right away? There is plenty of space inside the roll for a high-powered bug, and a stack of C or D cell batteries can be wired together in series parallel to power it for several weeks. Series and parallel are illustrated in Chapter 5: The Secret Lives of Batteries. Place some foam rubber around the batteries to keep them from falling out. Even better would be to place the prints inside a cardboard or plastic mailing tube. Remember that the microphone has to be placed so that it will be able to pick up sound.

Sometimes you may need to get in, but only for a few seconds. What's another thing that is likely to be found there? A lunch bucket someone left behind. There is a lot of space inside a Thermos jug. You dress for the part, walk in, and ask if they are hiring. Then forget your lunch bucket and the drop is made. Just in case they are hiring, it helps to know something about the trade. Have a good story already made up. Know the lingo. They may (probably will) tell you that they do not hire at the site and that you will have to apply at the union hall, so you will be expected to leave immediately.

Diversionary Tactics

I believe it was in a Robert Ludlum novel, where someone was teaching someone else how to think like a spy in the field. The situation is that said spy has become trapped in a railroad car. The enemy in front of him, the enemy behind. What to do? Attract attention. Feign a sudden and very painful illness. Moan and groan. Collapse onto a seat, gasping for breath. Everyone is looking, and the enemy is less likely to make a move.

Here is a situation where you need to create a diversion. Accidentally bump your knee into something, and, while you are cussing at your own clumsiness and the pain, slip the lunch bucket under a table and, apologizing for being such a klutz to keep them from saying anything to you, slip out the door with your hands in front of you. The same thing can be done in a lawyer's office. Remember the "inventor"?

Use your imagination! Visualize! Distract!

LET GEORGE DO IT

Maybe you won't need to get inside at all. Suppose you knew that the project foreman was going to be out of town for several weeks. Arrange to have a courier deliver the blueprints inside a mailing tube, with a note that they are confidential and are to be opened only by the foreman when he returns.

Find another way to prevent anyone from looking at the prints. Call them and explain that the messenger delivered the wrong ones, and he will be back to pick them up Monday morning. By then, they probably will have been forgotten. Play it right and you might be able to work this scam several times, replacing the roll of prints when the batteries start to discharge.

LENGTH OF SERVICE

How long is the device to be used; how long do you need it to transmit? You already know the battery life of the transmitter to be used (don't you?), so if the situation calls for long-term surveillance, then you need to arrange for a constant supply of power. Going back to the scene to replace a discharged battery is a very bad move. Again, case the target if possible. What is there in the area that plugs in and could be replaced with another, similar one without it's being noticed? In a lawyer's office that has Tiffany lamps, perhaps nothing. Look around and see what's there.

When constant power is needed, the transmitter may have to be installed inside a wall plug (not much space there) or lamp, clock, etc. Usually. There are exceptions. There may not be enough time to pull the cover off a wall outlet; you might get only one chance, and you may not have any other options for concealing the transmitter. So you work with what you have. Now suppose there is a sofa against a wall in a small, crowded office. There is no other place the sofa could fit, so it isn't likely to be moved. And—oh lucky you— there just happens to be an electrical outlet behind said sofa. Observe: How clean is the office? Does it look like the cleaning lady has nothing else to do but dust and vacuum? Or is it rather dusty, especially behind the couch? If so, then you might just plug it into the outlet and it might not be noticed for a long time. Maybe.

If the estimated length of surveillance needed is, in your judgment, less than the time you think there is until the place gets cleaned again; if you have very little time to make the drop; and if no other constant source of power is available, you can take a chance and plug the power supply into the outlet. Using an adapter modified as in the previous chapter will reduce the chances of its being seen.

If the expected length of service is a short time, such as a particular event, meeting, etc., a battery-powered unit may do the job. However, can you be sure of the time period? People call in sick; meetings are postponed, etc. Remember, you may get only one chance.

Is there a place where you could hide a larger, longer-lasting battery without its being discovered? Perhaps inside of the sofa. Do you have an X-ACTO knife, needles, and several colors of thread in your spy kit? Are there books that could be substituted with one that is hollowed out? Also, keep in mind that there might be a professional sweep done periodically. Typically, this is once or twice a year, but it may be more frequent. The White House, so I hear, is swept every day.

One quick method, for short-term surveillance, is to place the bug inside a candy bar wrapper or empty cigarette pack and punch a few small holes for the sound to get in. Then toss it in a wastebasket. This is one of those rare instances where you might get the transmitter back. Hire a wino to go Dumpster diving for you.

MORE IDEAS

Surveillance transmitters will probably be noticed unless they are either hidden out of sight or placed inside something larger that can be in plain sight. Remember PK Elektronik? They make bugs built into all sorts of common household and office objects.

Now, suppose the person you want to listen to spends most of his time in the garage restoring a '57 Chevy Nomad, a most worthy and honorable avocation. One of the things that will likely be lying around is a car battery. Or several. A car battery (assuming it is charged) will power most surveillance transmitters for many months. The transmitter could be taped to the side of a battery that is on a shelf or half buried behind old fenders and radiators and likely not be noticed for a long time. Some spy shops sell car batteries, realistic-looking ones, that have been hollowed out to use as secret hiding places. There is enough space inside for a large bug and a motorcycle battery. Would an extra car battery be noticed? Could you make it look like a used one and then switch them?

A variation on this theme is to build the transmitter into an old voltage regulator. Lay it beside the battery and use fine wires to connect it. Smear a little grease over the wires to hide them. If the battery is moved, the wires will likely come loose without their being noticed, or with little attention being paid to them. For the antenna, use a piece of wire of the type that is usually connected to voltage regulators, left in place.

You remember about polarizing, don't you?

IN BETWEEN

Remember the blimp hangar? Anything between the transmitter and receiver, particularly metal objects like desks and filing cabinets, will block (absorb) or reflect part of the signal. In the ideal situation, the antenna will be in a direct line of sight to the receiver, with nothing in between. While this isn't always possible, it is the next consideration. You work with what you have.

While deciding where to hide the transmitter, keep in the back of your mind the direction of the listening post. Observe the different parts of the target area, and note what is there that could interfere with the signal. Are there other rooms in between? Is there a metal garage door in between? Does the building have aluminum siding? Is there a power company substation between the building and the listening post? Think about these things before making the final decision.

POWER

In any situation, it is a good idea to use only as much power as is necessary, with a margin of error, and no more. If the transmitter has to reach a listening post that is only a hundred feet away, there is no need to use

100 mw when 25 will do. The greater the range, the more likely the transmitter will be heard by someone other than the intended receiver. Particularly if it transmits on the FM broadcast band. Why take unnecessary chances? Field-test if possible.

ANTENNAS

Use what you learned in the previous section on antennas and apply it with a little imagination. In a pinch, literally anything made of metal (that is not grounded) can function as an antenna. In some situations, you will have very little time and will have to act quickly. Remember about polarization. Have an alligator clip already connected to the antenna wire. You can clamp it onto a curtain rod in an instant. Use a file to sharpen the teeth of the clip so it will cut through the paint and make a good connection. It can also be used to clip a wire antenna to the inside of drapes.

PROBABILITY OF DISCOVERY

A surveillance transmitter may be discovered the instant it is turned on. The offices of some politicians, big businesses, and law firms may have RF detectors permanently installed in sensitive areas. Telco switching offices, research labs, government installations, and the like are also likely to detect surveillance transmitters instantly. Remember the "inventor"? Mounted under the lawyer's desk there might have been a high-quality field strength meter, such as the one made by Marty Kaiser. And you have been caught in the act. Now, the odds are that this lawyer isn't going to let you know he knows. He probably won't say anything. Rather, he will have you followed to see who you really are and why you want to bug him. Anytime you install a transmitter, you take that chance.

This is a good application for a remote-control transmitter; turn it on after you have left the office. Another trick is to use a timer to start it operating after you are long gone. Such a timer can be built with a 555 chip and a handful of resistors and capacitors. Many places that don't have detection equipment installed will have technical surveillance countermeasures (TSCM) teams sweep the area periodically. So much for long-term surveillance. That's another chance you take. Employees of such businesses may also have detection equipment installed in their homes and have a sweep team come in periodically.

On one end of the scale are "just ordinary" people. Much more vulnerable, but of little interest to big business and government, they are unlikely to be bugged. If anyone were to bug them, it would probably be a domestic husband-wife thing—one person suspicious of another. People are becoming more aware of not just surveillance equipment that is available but also the countersurveillance equipment that is affordable. The probability of discovery increases every time someone reads a book like this one and does something to prevent themselves from becoming another statistic, another Quiet Victim. People's lives can be devastated because of electronic surveillance. The more they realize this, the more likely they are to be prepared.

LOCATION OF THE LISTENING POST

Where the listening post is to be located is another factor to be considered when deciding on the type of transmitter, the location, the antenna to be used, and so on. If the transmitter is to be an alarm system in your office and your home is across the street, most any of the medium-powered types 25 MW or so, should work fine. If the distance to the target is several miles or more, then you have a different set of circumstances to deal with. Reliable surveillance over such distances is not so easy to establish. (More on this later.) So, here you have another decision to make. Either you can set up a listening post closer to the target, or maybe increase the power to overcome the distance problem.

There are several ways to set up a listening post close to the target. You can rent an office or apartment within range, use a vehicle parked nearby, or hide the listening post inside something.

OK, time for another one of my guessing games. What the hell could someone possibly hide a listening post inside of? A storage locker for bicycles. A lot of people are riding them these days (except in California, where people seldom get out of their cars), and they need a place to park them. Public transit park & ride and train stations have bicycle lockers. There is enough space inside one of these for the receiver, a modified long play recorder or repeater transmitter, and 42 years' worth of batteries. Also, they are usually made of fiberglass, which won't block the signal like steel would.

Is there a park nearby, within range? Or a bus stop? Good for short-term monitoring.

Another trick: n any metropolitan area, there are messengers, couriers who deliver small packages to law offices and other businesses. They are so commonplace that, as mentioned above, no one pays any attention to them. Sometimes, when business is slow, they may be found sitting in front of an office building for long periods of time. They have bags or backpacks for their deliveries (perfect for a tape recorder, video cam, and whatever else) and they carry radios. An excellent "listening post" for part-time long-term surveillance.

WHAT TYPE OF TRANSMITTER TO USE

After you have thought about all of this, you can decide what kind of transmitter to use. Of course, there is also the matter of cost to be considered. In any case, a well-equipped spy will have several different models in his bag of tricks. Since you never know what you might encounter when you go in, what might have changed since you did your recon, it is wise to be prepared. Use the best one you can afford to lose.

RECOVERING THE TRANSMITTER

In a word, don't. When you install a listening device, you should write it off. You don't get it back. Returning to the scene is not a great idea. Someone might have discovered it and be waiting for you. Sometimes people make the mistake of telling someone else that they have installed a listening device. Then they start to worry about it and decide to remove it. Don't do this.

Sometimes people use a very expensive transmitter and don't want to lose what they paid for. So they go back to get it. Stupid.

Someone who has been in The Biz for quite a few years told me that this does happen now and then. He has seen where something has been removed from a phone, for example. The wires were cut rather than disconnected, or sometimes even ripped out. If the person did the job right, there would be no reason to go back to get it, especially if he knows it has been discovered. And if he did it right, he would have used a device that can never be traced back to him.

Don't get yourself into such a situation. It ain't worth it.

"... Yeah, it cost me a bundle, and it's probably still there. Go get it? Hell no ..."
—Comment made by someone who used to work both sides of the fence

KNOW AND TRUST YOUR EQUIPMENT

SCUBA divers and trapeze artists learn to know their equipment well, keep it maintained, and trust it. Their lives depend on it. While the life of a spy may not depend on the listening device used, the success of the operation does.

Before you buy any transmitter, find out as much about it as possible, and after you buy it, test it. Use different types of antennas and place it in different locations. Use different receivers to check range and reception, and have someone stand at different distances from the microphone to check out the audio quality. Measure the battery life.

Check it for harmonics. A harmonic is a multiple of the actual transmitting frequency. If the transmitter is set on 100 MHz, there will harmonics transmitting on 200, 300, 400,etc. MHz. It is possible for these signals to be received on a TV set. The best way to check for harmonics is to use a spectrum analyzer, but if this is not possible, use a TV set, one that has an earphone jack if possible. Turn on a radio and place it close to the transmitter to give it something to send. Then set the TV on each of the channels that are used in the area, one at a time, and tune the transmitter through its range. Listen for the sound of the radio, but also watch the screen. Look for any kind of interference. This will probably be what is called the Herringbone Effect, which is a series of curved, closely spaced parallel lines that float across the screen. Turn the transmitter off and see if the lines go away. If so, you have what is known in The Biz as a dirty bug. One that has strong harmonics.

9
Receivers

When you come right down to it, a transmitter is a transmitter: no matter what type or how fancy or exotic or expensive, it sends out a signal of a certain intensity or strength. That signal will radiate a certain distance and may be received, intercepted, within that distance. But what is used to receive the signal, the radio, can be just as important as the transmitter. *Maximum effective range.*

Generally, the better the radio the better the reception. However, a fair receiver with a good antenna will pull in the signal better than a high-quality radio with a poor or no antenna. While testing some transmitters for the first edition of *The Bug Book,* I used a PRO-2006 scanner and ICOM R-7000 communications receiver and an old beat-up Sony portable FM stereo. The Sony, with a telescoping antenna, could hear one of the wireless microphones when the scanner, with no antenna, could not. With the same antenna the scanner could pull in the signal from more than twice the distance as the Sony. Use the best you have available.

COMMERCIAL FM RADIOS AND STEREOS

If you are using an FM receiver in a fixed-location listening post, your choice of antennas is greater. If an outside TV antenna is available, try it.

Otherwise you can use a folded dipole cut to the right length, or maybe a longwire. Sometimes the listening post will need to be a park bench to be within range of the transmitter. Use the best radio you have available, and remember about polarizing the antennas. Your listening post may be a vehicle, and while car FM radios work fairly well for this purpose, a mobile or portable scanner will work much better. A magnetic mount antenna will work better than the car's antenna, especially if you cut it to the right length. If the transmitting antenna is horizontally polarized, try stringing a wire or dipole across the rear deck, or conceal it inside a rooftop luggage rack.

SPY TRICK

When using a stereo receiver to monitor an FM bug, put it on mono and disable the mute control.

Modifying FM Receivers

Most of the inexpensive bugs and wireless microphones operate on the commercial FM broadcast band, which is 88 to 108 MHz. Some are limited to this range, but others can be adjusted to transmit above or below. If you have such a transmitter, then it is a good idea to use it outside the FM band, since this reduces the chances of some "unauthorized" person hearing the signal. So, what do you use at the listening post to hear it? If you have a scanner or communications receiver that has wideband FM mode, fine. But suppose you do not. Some FM radios can be modified so that they will receive above 108 or below 88, which is called out of band, or OOB (not to be confused with Out Of Body, which results from watching too much television). It will work with any radio that has manual tuning (i.e., that has a knob you turn to change stations). This includes both solid state and the old tube types, should you have one. This can also work with electronic tuning, where you enter the frequency on a keypad, but is a tad complicated and is not discussed here.

Before you start, be aware that making the adjustments for OOB reception might decrease the sensitivity, and it will cut off reception at one end of the FM band. So, may I suggest that you use a receiver you can afford to lose in case you should damage it. It is also possible that this modification will decrease the sensitivity of the radio slightly, unless it is realigned after the adjustment. If you use it to listen to weak FM stations, you may no longer be able to receive them.

A Zenith transistor radio from the late fifties. I found it at a sidewalk sale for two bucks, and it still works. The clear plastic box at the top is the variable capacitor, showing the two screws that are adjusted for out-of-band reception. As this is an AM radio, there are only two. An AM-FM would have four.

First, disassemble the case, remove the back cover, whatever it takes so that you can get to the insides. The component you are looking for will be in a small box, about 1 1/2 inches square and 1/4 inch thick (it may be smaller or larger, depending on the overall size of the radio).

NOTE: Either type of radio will have a number of small metal "cans," about 3/8 inch square, soldered to the printed circuit board. They have one hole in the top and an iron ferrite screw that requires a hex-shaped tool to turn. These are intermediate frequency (IF) amplifier tuned circuits. They are not what you want to adjust. Doing so will throw the receiver out of alignment and may ruin reception. Realigning them is possible, to some extent, through trial and error, but doing it right requires a signal generator and other test equipment.

The component you want to adjust, called a variable capacitor, will usually be inside a little box made of clear plastic. As you turn the tuning knob, you will be able to see something move inside.

On the top of the box will be two to four small, flat metal screws. One of these will be adjusted to make the modification. You might make a mark on the plastic to indicate the original position, in case you later want to reset it. This adjustment can be made at either end of the FM band: above 108 (which is in the aircraft navigation band), or below 88 (which is TV channel 6).

Tune the radio to a strong station that is very close to the end of the band you want to use. Now s-l-o-w-l-y turn one of the screws and listen to see if the station starts to fade. If it does, you have the right one. If not, turn it back to where it was and then try the others. When the signal becomes weak, but you can still recognize it, adjust the tuning knob so that you are once again receiving it as loud as before, and then note the dial indicator. What you are doing is sliding the radio's 20 MHz coverage slot so that it will now receive above 108 or below 88 MHz. Continue this, noting the new dial reading for the station, to determine that you are moving in the direction you want—higher for the top end of the band and lower at the other. The actual reading makes no difference; it is the distance it has moved that counts. You want the shift to be at least 3 MHz because some FM radios already will tune 1 or 2 MHz outside the band, and you don't want them to be able to hear the transmitter. When you have moved it that far, fire up the transmitter and adjust it to the frequency the radio is now tuned to. The last step is to take a different (unmodified) radio and try to tune in on the bug's signal. If you don't hear it, then the adjustment was done right. This completes the modification.

Tailor Made

If you want to buy a radio that has been so modified, C-Systems has the RX1 for $92. The frequency coverage of the FM section of this receiver is retuned to the bottom end of the aircraft band (108-136 MHz).

USING A SCANNER

Scanners have much better sensitivity and selectivity than FM radios and will have better reception when used with surveillance transmitters. The sensitivity of a scanner might be less than one microvolt, where an FM radio is something like 15 microvolts. This makes a very big distance in the maximum effective range. There are dozens of scanners on the market, and virtually all of them cover the 400 MHz, area where most crystal-controlled wireless microphones operate, as well as the commercial FM band. You can spend a lot of money on a scanner, but it is not necessary to buy a 100-channel type to listen to one frequency. Pawn shops and second-hand stores often have 16 or 20 channel scanners for a hundred bucks or less.

COMMUNICATIONS RECEIVERS

The difference between a communications receiver and a scanner used to be that the former had a tuning knob and the latter had keyboard entry only. Now, many of the better scanners have manual tuning knobs, but whatever the difference, communications receivers are usually better built, more expensive, and have more options and accessories.

There are dozens of brands and models to choose from, Kenwood, Yaesu, and ICOM being the most popular. I am partial to ICOM; I once owned an R-7000, which is an outstanding radio, but the others are also excellent. Sometimes they can be found at ham radio swap meets or through any number of publications for and about amateur radio. Most of them have both wide- and narrow-band modes for commercial FM band transmitters, but their most important feature is that you can tune them exactly to the bug frequency by just touching the tuning knob. With a scanner you have to enter the numbers on the keypad, which takes longer.

HIGH-FREQUENCY RECEIVERS

The HF band is from 3 to 30 MHz, and there are many brands of radios that cover it. Sangean has a number of portables that receive FM and short wave and range in price from $100 to $300. There are also Lowe, Sony, and others. A little higher in price is the ICOM R-71A, a classic time-tested radio that goes for about $1,200. Check prices at Ham Radio Outlet (listed in Appendix C). Watkins-Johnson makes an HF receiver that is every short-wave listener's dream. The HF 1000, described as the best there is in the Grove Enterprises catalogue, will set you back $3,800.

The ultimate radio is the ICOM R-9000. It covers DC to infrared, all modes, and a built-in computer with monitor. About $5,500. Most dealers sell only to "FCC approved" personnel.

Whatever you use, try to get the best you can afford without spending more than you need to. Read the specifications and operator's manual if possible, or ask questions on the Internet.

10
An Exercise in Surveillance

OK, troops, let's take what you have learned and put it all together. We will start with a rehash of some basic stuff, and then you will take over. You will take a hypothetical target area and see how you might go about bugging it using an RF transmitter. As you know, there are many things to consider. Every situation is unique, but they all have in common the principles you have read about so far. Power, antennas, acoustics, listening post . . .

Decisions, decisions. Of the many decisions to be made, the first one is how much time is available?

RECON REVISITED

If possible, in the time I had, I would make a profile of the target area based on all available information. The more time available, the more time I have to prepare for the operation. Sometimes, there is little to research, little to consider; sometimes, much has to be done before the drop.

You want to get some information from a certain person, and you know a little about his habits. He goes to a particular restaurant on a regular basis. So you decide to hide a transmitter in the restaurant. The mechanics of the operation are simple. A medium-power unit stuck under the table with the two-sided adhesive that you always keep in your spy kit. The listening post is a van parked across the street. A simple enough operation with a few variables.

Which table? How many tables are there? How many transmitters do you have? Is there time to observe, to see if the subject has a favorite table that he always uses? Is there an employee that will provide this information? Will that employee install it for a price? Will he keep his mouth shut or will you get another invitation to go swimming in the Hudson River? Can you think of a way to cause the person to take a particular table? Could you reserve a number of particular tables, under different names, to reduce the number of tables left available? Can you arrange for one or more of the restaurant employees to miss work that night—and for someone who works for you to just happen to go to that restaurant asking for a temporary job?

Everything you can find out about the target area—the people who are there, when they are there, the type of operation it is, security, chances of discovery, probability of obtaining the wanted information, access to the area, acoustics, equipment indicated, equipment available, location of listening post—all these things should be thought about. Almost anyplace can be bugged. Some are easy, and some are quite difficult. It depends on who is doing it. Some places are a real challenge, such as the operations room at a Strategic Air Command base; the National Reconnaissance Office "Blue Cube" at Moffett Field Naval Air Station; the Trilateral Commission offices (which are at 345 E. 46th St. in Manhattan in case you're interested); and, of course, many of the three-letter federal agencies.

ON THE FARM

This is gonna be a hard one, requiring a fair investment in both time and equipment. I am using this for my usual reason: to get you to consider all possibilities, use what you have learned so far, think and visualize. If I made everything easy, you wouldn't learn very much, eh?

Your objective is long-term (longer than a battery-powered transmitter would last) room audio surveillance of the living room in a farm house, located five miles from the nearest town.

"Five miles?" you ask.

Yup.

"Oh, shit," you say.

Yup. And now, you take over. This is your operation.

More Recon

This may be fairly easy, if you go about it the right way. One secret to casing anything (yes, I have said this before) is to look like you belong where you are, and to always have a reason should someone ask. Such as someone who is . . . well, have you ever noticed how big the barrel of a .357 is when it is pointed at you? This is the ultimate test of bladder control.

By pretending to be lost and stopping to ask directions, you can get a look at the general area. A vidcam concealed in the car will make a permanent record for later review. How many people are there at the time? Is it a large family? How many dogs are there? A "teacup" poodle isn't much of a threat, but suppose they have a 240-pound English mastiff? What kind of locks are on the doors of the house? Is there an alarm system? If so, make a note of the type and the company that installed it. Does it have a basement? How many other buildings are there? How far are they from the house? Do the doors face away from the house? Are they locked? Are there power lines leading to them? Is there a wooded area near the house to provide cover when you go in?

To get a look inside the house, you might pretend to be a salesperson of some kind. An advance call or two might provide some insight here. You might be invited to come out and demonstrate a software program that is custom made for the many things a farmer has to keep track of. If that doesn't work, you might read up on some subject that is of interest to a farmer and interview him for an article you are writing. Pretend that you came into some money and always wanted to have a small farm, and need their advice. Pretend to be a genealogist searching their family name.

Find out what you can about this family. Search public records (without leaving a paper trail) and try to find some useful information that you can tailor to your needs. Do whatever it takes to get in the place to have a look.

While inside, you would be making mental notes as to places the device could be installed. They might have an old tube-type radio in the room—a perfect place for a large, high-powered bug, and a constant supply of power. You could observe it and mention that you have a similar old radio that you were planning to give to Goodwill. Which they can have for free. (As soon as you have built a transmitter into it.)

They might have a large hardwood china cabinet, which is too heavy to be easily moved. Between it and the wall there might be space for the transmitter, and a wall outlet in the baseboard. Using the type of adapter described in Chapter 5, it would provide power and not be easily noticed.

The Listening Post

A mobile listening post is probably not a good idea because the vehicle would sooner or later be noticed. Probably sooner. Even if a different one were used every day, it would still be noticed, and someone will eventually come around to investigate.

A hidden listening post is possible, as long as you can find a way to access it to change the tapes. However, in this case the usual long-play recorders that are good for five or six hours just won't do the job. You would attract too much attention by going there several times a day. There are commercial digital recorders available from CSE Associates in Maryland that have much greater capacity, but they are large, expensive, and require 110-volt power. And even these have to be maintained.

So you decide to set up in the town that is five miles away. Rent a room or apartment, maybe even an office, that is in a direct line with the target (if possible) on the highest floor available. The receiver will be an ICOM with a preamp. Depending on the frequency to be used, you might have a beam (yagi or log periodic) or a longwire antenna strung around the room. It would be better if you could string it from an upstairs window to a tree or whatever was handy, but would that make anyone suspicious? Would they buy a story about being a ham radio operator or short-wave listener? Could you come up with something else that they would buy? It is little things like this that can make the difference between success and failure.

The Transmitter

The next problem gets into the old argument about range. Some people will say that there is no way to get a surveillance transmitter to work over such a distance. Wrong. Some will say that it is damned difficult. Right. There are several possibilities here, all of which can be used to transmit the signal five miles:

1. Use high (VHF) frequency, a lot of power, and an antenna as high in the air as possible.
2. Use a small bug with a repeater.
3. Use a spread spectrum transmitter.
4. Use low frequency, less power, and an antenna closer to the ground.

Let's look at each option.

1. *High frequency.* If there is nothing to block the signal path, this will work using five watts output or less, but it will require constant power. The antenna will have to be located on top of something—one of the outbuildings, the barn, possibly even the house. Access may be a problem. And there is the transmission line leading to the antenna. Is there a way to hide or disguise it? Could it be concealed behind a drainpipe? With VHF there is a fairly high probability of discovery: even in a rural area someone might well be listening. You would have already tried to find a frequency that has no activity over a period of time (if you had enough time), but it is possible that someone, a "kid with a scanner," might overhear the bug. And the FCC might also intercept the signal. They have sophisticated monitoring stations at different locations, as well as mobile units.
2. *Small bug with repeater.* This eliminates the transmission line, but you still have to get power to it, and the probability of discovery is unchanged. The power is the same, as is the frequency.
3. *DSS transmitter.* This is better, as it virtually eliminates the possibility of the signal being intercepted and reduces the amount of power needed. One or two watts should be plenty. The antenna doesn't have to be at as high an elevation, but constant power is needed, and there is the transmission line to deal with.
4. *Low frequency.* Many problems are eliminated here. One of these is the cost; the transmitter can be made from an old tube-type radio, and as you read earlier, it can be "tweaked" so it transmits below the AM broadcast band, making it less likely to be discovered. The antenna can be located close to the ground, eliminating the need to climb up on a roof, and probably making concealing the transmission line easier. It will also be easier to get power to the equipment.

Now, where can you install the transmitter? Option: inside the house.

The AM transmitter will be about the size of a breadbox, making it a bit difficult to hide and still more difficult because it requires constant power. It is unlikely that you will find such a location. Even if you have had plenty of time to case the place, this is improbable. Oh, you might find a place, but there is still the antenna to consider. At low frequency a longwire is indicated and stringing it around the living room is not specifically a good idea.

There might be a good place in one of the upstairs rooms—perhaps a bedroom that is no longer used. Space and power might be available, and there might even be a way to run the antenna wire out a window and along the side of the house, concealed as best you can. But this creates more problems. You would need a ladder to string the wire, and you still have to get sound from the living room microphone to the transmitter.

Another possibility, a bit of a longshot but . . .

You are starting a new company that installs burglar alarms. In order to get established, offer them a system for free. If they already have an alarm system, convince them that the one you offer is, in some way, better. Or, why not have a second, backup, system as long as it is free?

This is gonna require that you make up some "official"-looking forms, and there is a very good chance these people will do a little investigating of their own. Do you have a business license? Where is the shop? Would a sympathy ploy work? "I'm starting out on a shoestring and working out of my van to feed my 16 children . . ."

If they decide to let you install the system, you can place the transmitter inside an authentic-looking alarm box. Contact microphones disguised as glass break detectors will be placed on the windows, taking care of the problem of getting audio to the transmitter. But what about the antenna?

....

Pop Quiz
Could I use the cables that go to the magnetic switches as the antenna. Why not? Or, is there a way to improvise? Think about it.

....

Again, this is not the best way to go about setting up surveillance, since it is very time-consuming and expensive. And, of course, they will be able to describe you to a "T" should it become necessary. If this were an assignment hundreds or thousands of miles from where you live, it isn't likely to be a problem, but in any case effect some sort of disguise (beard/no beard, hair color, etc.) and be sure not to leave fingerprints.

Another possibility: There is a storage shed 55 feet from the house. It is secured with an ordinary Master padlock, which is easy to pick. It has power coming in and uses old white ceramic knob and tube wiring.

Inside the shed is a long, high shelf. It holds several rolls of wire fence laid end to end. They have been there for a long time; the dust is very thick and shows no signs of having been disturbed. Also on the shelf is an old galvanized sheet metal box made for some unknown purpose.

The transmitter can be placed in the metal box, and be connected to the power line by splicing into the wiring behind two of the knobs.

For an antenna, there are several possibilities. It would be possible to stretch a long wire on the back of the shelf, but this could disturb the dust, and someone might notice it and investigate. Another idea is to connect the rolls of fence wire to each other with short jumpers that have alligator clips with sharpened teeth. The antenna wire is connected to the roll closest to the steel box. If the transmitter frequency is variable, a grid dip meter might determine the resonant frequency of this makeshift antenna. This is iffy. A better way is to measure the length and make some calculations.

There also might be a wire stretched from this building to another; perhaps an old signaling device or intercom, even a phone line that is no longer used. A near perfect antenna, if it faces the listening post. It's not practical to resonate something this long, but you need to know the precise length. A tape measure could be used but there is a better way. Easier and much faster. Can you think of it?

OK, you decide that the transmitter will be set up in the shed. Now how do you get the room audio to the transmitter?

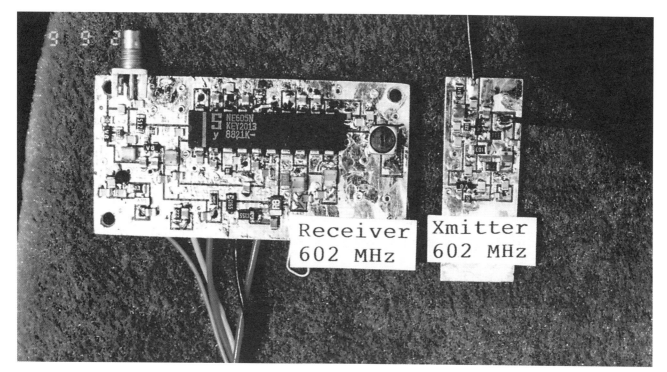

A surface-mount UHF transmitter. Size is 14 x 36 mm.

1. Use an infrared transmitter.
2. Use a small RF transmitter.
3. Hardwire the microphone.
4. Use the phone line.
5. Use a subcarrier transmitter.

Let's look at each of these options.

1. *Infrared transmitter.* Infrared is counterindicated for a number of reasons. This is a line-of-sight device. The audio is coming from the living room, which faces the front of the house, and the shed is in the backyard. Now, even if you were not faced with this problem, remember that since the transmitter has to be visible to the receiver, it will likely be visible to the people who live there. Then there is the need for constant power. Lastly, 55 feet may be too far for the readily available commercial systems. So, you would have to build a better one. Much time and expense involved here.
      Infrared is definitely out.
2. *Small RF transmitter.* This might be intercepted. Slight chance, but someone living there might have a scanner and stumble across it. In this situation, a bug that works in the high UHF area, such as the one pictured here, would be better than most other frequencies. But it needs constant power. One of the transmitters Sheffield used to make uses the phone line for power, has a wide frequency range, and transmits room audio when the phone is not being used. If you can come up with a similar transmitter (and I don't know who makes them), this is also good.
3. *Hardwiring the microphone.* This is probably counterindicated. Remember that the living room faces the front and the shed is in the back. So, you would have to run the cable through the house and then through the yard. Concealing it would be very difficult.
4. *Phone line.* This is a possibility. If you can get access for long enough, a microphone could be placed

inside the connection block on the wall. Normally there are four wires, two of which are not used. You remember which colors they are, don't you? These two wires will lead to the SPSP or network interface on the side, or more likely the back of the house. From there, the line might go to the storage area for an extension, which may or may not still be connected. If not, a line-powered RF transmitter could be used.

5. *Subcarrier transmitter.* Since there is no way to avoid the need for constant power and you have to plug in whatever is used, this is the logical choice—the signal is transmitted through the power lines and is very unlikely to be intercepted. But just to be sure, you have a subcarrier current detector with you and use it to check the line for an existing device and to make sure it works.

This brings up another point. About making people suspicious. You might be thinking ahead and wondering if maybe the people there have a subcarrier "wireless" intercom that would pick up the audio from the one you plan to use. This probably won't be difficult to determine, if you call them and ask something like, "We have a new model intercom that's better than the one you bought from us last year, and we thought . . .

"What? We have no intercom here. Are you sure you have the right number?"

In order to get in to set it up, you might create a diversion by mailing the farmer a gift certificate purchased from a nice restaurant. Reservations for the entire family would be made for the night you want to go in. To avoid making them suspicious, you might send them a note saying that you were once lost and stopped for directions. Because of them, you were able to make an important

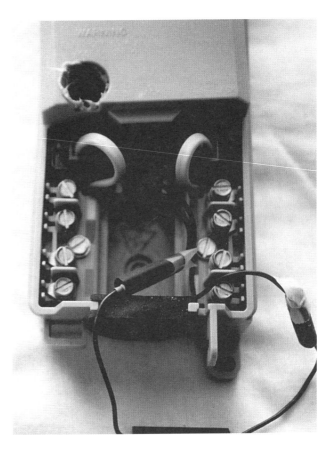

As I have said elsewhere, a good way to prevent someone from finding a surveillance device is to conceal it in something else. Taking this a step further, keeping anyone from opening what the device is hidden inside of may further decrease the chance that it will be discovered.

This network interface had the Torx screw slot drilled out. It was opened by melting a hole in the cover with a soldering iron. Inside was a small series transmitter, brand unknown. You can see the edge at the bottom of the photo. This bug was found because it malfunctioned, making the line go dead and causing someone to investigate.

appointment on time that otherwise would have been missed. This is your way of thanking them.

The possibilities here are endless. You could have been a sales rep for a marketing company in a faraway place that would give them a new TV if they would agree to monitor certain commercials and write a report on them, that was to be sent to a mail drop once a month.

You could have been an itinerant fix-it person who came by looking for something which needed repair. A magnetic sign on the door of a rented pickup truck could identify it as the Acme Furnace Inspection Company, who does free safety inspections.

You might have noticed something that plugs in, and accidentally knocked it over and broke it. You apologize profusely, and insisted on personally taking it out to be repaired.

You might have sent them a smoke detector as a gift. One that plugs in, so they never have to worry about the battery going dead.

Think. Visualize. Improvise.

While building a transmitter into something that plugs into a wall outlet will guarantee constant power, there is no guarantee that the subject will plug it in. Send them a lamp they don't like and it will probably end up in the basement. What would you do with an ugly lamp?

Aarrgghhh: Another Pop Quiz

There were three rolls of fence, each 6 feet in length. Including the jumpers the overall length is 20.5 feet. What frequency would be best, within the range slightly below the AM broadcast band, and what fraction of a full length antenna would this constitute?

MISTER MURPHY

Finally in our exercise, let us throw in a few monkey wrenches—the sort of things that spies may encounter in real life.

Q: Remember about dogs, and that dogs sometimes try to eat spies. Well, OK, at least chew on them for a while. What do you do about that?

A: You take a tranquilizer dart gun. A big one for the Mastiff, a small one for the poodle.

Q: Before I get into the house, the people come home unexpectedly, or a neighbor drops by. What do I do?

A: Normally, you are always better off to work alone. But in an operation like this one, it is a good idea to work with a partner. They will be watching out for anyone who might approach the target area, and warn you by two-way radio. If you are working alone, you can set up a perimeter alarm. There are many ways to do this, and you can get some ideas from Shomer-Tec, listed in Appendix C.

Whether you operate with a partner or alone, always have a cover story ready. At least one. And a few props. A pair of coveralls with an official-looking patch from the local power company is a start. Again, the magnetic sign on the car door adds to the credibility. You are looking for a problem in the transmission lines from the [insert name] power substation, and you whip out a field strength meter or other electronic device and start looking like you are doing something official and important.

Q: I am actually inside the house when they come home. How the hell am I gonna explain that?

A: Again, if you work with a partner, this will not happen. Otherwise, you'd damn well better have a good story ready. Believe it or not, this one has actually, in real life, been done.

Someone I know personally broke into a house in a rural area. This was not to install a listening device, nor was it to steal anything—it was for something else. And while the people were a bit skeptical, it worked. As soon as he got inside, he did several things:

He located a sheet of paper (that he did not bring with him) and wrote a note on it. This was folded over some cash (that he did bring with him) and laid on the kitchen table. Then he opened a container of sugar, made a point of spilling a little on the counter, and mixed some of it with tap water in a drinking glass.

The note explained that he was a brittle diabetic. (I don't know if that is a legit medical term or not.) His car broke down, he had forgotten his medicine, and if he didn't get some sugar in his system he might die. The money was to pay for the window he broke to get in. As this was a house in a rural area, miles from a store or restaurant, like in our exercise, it made enough sense that he was able to get away without ending up in jail for B&E or a hospital emergency room for BR (buckshot removal). A medical ID bracelet would have added to his credibility.

If you can damage something in the target area, such as a typewriter, it may have to be sent out to be repaired. If it has a label with the name of the company that sold it, you know where it will probably be taken. How much security did you ever see in a typewriter repair shop?

Or you might have had an official-looking ID card and badge in your spy kit, "proving" that you are an agent of the National Bureau of Intelligence or some such fictitious agency. The agency traced a vicious cyberterrorist to the area, and you are out of radio range and had to find a phone to call the field office. Put them on the defensive; imply that they should cooperate in helping to find this lowlife creep. Tell them that you were not yet able to get through to your office in Washington, DC, give them a number there (one you know will not answer at that time of day), and ask them to call it. Ask them to advise the person who answers that Agent Donahue has located the RS232 suspect and urgently needs backup personnel. Leave a few bucks to cover the cost of the call and a business card if you have one. Then split. Fast.

If you get caught inside, you have absolutely nothing to lose by trying to bluff your way out. There is no excuse for not having a good cover story, complete with props, ready at all times.

Q: What do I do if the people don't believe me and shoot me?
A: You get shot. That's a chance one takes when they try to bug someone. And there are people out there who really will blow your ass away.

Q: What do I do if I can't get the bug installed in the farmer's house?
A: Retire and write books about surveillance. Seriously, there is always a way. It may take time, but in such a low-security situation, there is little excuse for failure.

Q: Couldn't there be an easier way to do it?
A: There may well be. A phone tap from outside the house would be much easier than breaking in, but remember the objective: room audio inside the house using an RF transmitter.

Q: Is there any guarantee that this is going to work?
A: There is never a guarantee. Anything could happen. But if you know and trust your equipment and set it up right, it should.

HOMEWORK 105

Now, let's go over some of what we have learned about planning a drop. Ask yourself the following questions.

Access

Can you get in to the target area?
Can you get in more than once without blowing the operation?
Is it possible to arrange for someone else to install the device?
Can you hide the device in something that will somehow find its way into the target area? (Remember about unwanted lamps ending up in the basement.)
Where, in the target area, will the transmitter be installed?

The Listening Post

Where will it be located?
Is a vehicle available that can be parked within range without attracting attention.
Is there a place to conceal the LP?

What time period is involved? Will you need constant power? Will a battery-operated device work, and, if so, are you absolutely convinced that you can get access in the time required? Will you have time to test reception and, if necessary, use an alternate listening post?

Equipment

Do you have everything you may need?

There is no excuse for botching the assignment. If you cannot get it done, don't take it on. Let George do it. Let him end up as food for fishes.

A few pages back I asked if you could think of a way to find the approximate resonant frequency of a piece of wire that might be used as an antenna. The answer is to use a laser measuring device. Stand at one end and shine the beam to the other end and read the display. Now that you know the distance, you can convert that to frequency.

STORIES: DID THE DEVIL MAKE THEM DO IT?

Just about everyone has heard about Jim Bakker. Many people have heard about the so-called spying that was going on in the PTL offices; Bakker was accused of setting up an elaborate spying system, which the media made a big deal of. Here is the true version of the Jim Bakker/PTL bugging story. This is from the actual report written by Kevin Murray, who did the sweep. It is reprinted here exactly as received.

In the spring of 1987 the PTL (Fort Mills, South Carolina) was under heavy public, media, and government fire. Jessica Hahn opened the floodgates of criticism by accusing Rev. Jim Bakker, et al. of sexual assault. Other close but not so loyal followers began to open their mouths too.

Even Georgiana Moss, wife of PTL's growth Messiah James Moss, pointed the finger, claiming Jim Bakker had wiretapped Moss's office telephone and bugged his car. The tapes, containing conversations between James Moss and his mistress, Donna Axum (Miss America 1964), were mailed to Mrs. Moss.

Embezzlement, fraud, wife swapping, homosexuality, insobriety, and extravagant living were a few of the other attacks in Bakker's personal wrestle mania. The public was no longer viewing PTL as PRAISE THE LORD, but rather as PASS THE LOOT. Jokes abounded.

There's a sucker born every minute.

P. T. Barnum.

1987 was not a good year for Mr. Bakker. He had fallen from grace. Harry Hargrave and Rev. Jerry Falwell took control. The IRS moved in, and, PTL hired MURRAY Associates, a counterespionage consulting firm from Clinton, New Jersey, to conduct a full-scale electronic eavesdropping inspection. The morning after the Bakkers were ousted, we were there.

The areas we inspected included the Bakker's 11-room (and 6-bathroom) apartment at the Heritage Grand Hotel and the entire top floor of the pyramid-shaped PTL Headquarters building. Both locations are on the grounds of the Heritage U.S.A. religious theme park.

Upon completion of our investigation (5/1/87), a brief verbal report was given to COO Harry Hargrave. A 10-page written report was sent the following week. On Friday (5/8/87) Harry Hargrave announced to the media that a radio transmitter had been found . . . and an elaborate bugging system had been found . . . capable of transforming the building public address system into a Big Brother-style listening post . . . activated by telephone . . . . [so that] you could listen to anything in the building . . . from anywhere in the world.

Everybody was shocked. If they were shocked, I was electrocuted. Here is what they really found.

PTL headquarters had a building-wide public address system. The microphone (and preamplifier) for this system was located in the corner of Jim Bakker's office, behind a pillar and some plants. It appeared that Bakker used this system to address the troops from the comfort of his office. On the day of our inspection Shirley Fullbright (believed to be Bakker's executive secretary) told me the system has been broken for over a year.

Our inspection, however, revealed the microphone and preamplifier were indeed on, and worked just fine. Room audio was being sent to the building's main-floor maintenance room. There the wires entered a wall-mounted junction box, marked "mike from J. B.'s office." The wiring then traveled a few more feet to the main building amplifier, which was operational, but was turned off at this time. It was not broken, as had been thought by Bakker and his top-floor staff. By attaching our equipment to this microphone line (from the main floor maintenance room) I was able to listen to all sounds coming from Jim Bakker's office (top floor). An old car radio speaker had been hung on the wall within two feet of the junction block. It also afforded a listening capability.

Telephone wiring junction blocks, with all the telephone lines for the offices upstairs, were located here too.

I advised Harry Hargrave that since this public address system was not in use, and its speakers were addressable on a zone basis, yet another method of eavesdropping existed (accomplished by attaching an amplifier to the speaker wires, thus turning the speakers into microphones). These speaker wires terminated in the building maintenance room along with Bakker's microphone line, and the telephone lines. All were easily accessible. In fact, from this secluded maintenance room one could easily listen to

- Bakker's private office, via the live public address system microphone located there.
- any of the telephones, by directly attaching to the junction blocks.
- any area of the building, by using the public address system ceiling speakers as microphones.

This combined with

- rumors of electronic eavesdropping at Headquarters, which were being fueled by accurate feedback.
- the old PTL leadership being falsely lead to believe the public address system was broken.
- the microphone and amplifier in Jim Bakker's office, which had been left on.
- a car radio speaker, with makeshift wiring, found at the focal point of all this wiring.

Suppression of our final report led us to believe, if any eavesdropping was going on, the eavesdropper had been listening to the Bakkers from the safety of the maintenance room . . . not the other way around.

This was my firm's involvement in the PTL case. I reported our findings in nonsensational, factual manner and was surprised to hear the alternate story released to the press. The FBI was also surprised by Mr. Hargrave's press account of the extensive bugging and they requested a copy of our report to review. They verified our findings through an independent investigation. No further action was taken.

Case closed. Another piece of history became warped. Jim and Tammy Bakker can be accused of many things. Harry Hargrave's accusations about an extensive electronic bugging system at PTL headquarters is not one of them.

11
The Phone Book Revisited

STORIES: AM I TAPPED?

Funny how things happen. I had intended to include info on the following device in *The Phone Book*, but at the time I did not have one on either of my lines. I had been meaning to set it up, just for the hell of it, but never seemed to get around to it, even though it takes only a few minutes. So I forgot about it until very recently. When I was suddenly reminded.

I first used this little goodie many years ago when most people had party lines. I sometimes listened to the girl upstairs talking to the old biddy next door about what a "weird electronic kook" I was (guilty as charged). But when I'd pick up the phone to eavesdrop, it sometimes made a click that they could hear. So they would change the subject. What I needed was a way to listen without picking up the phone.

If you have read *The Phone Book*, you may recall Betty's Bug. This was an incident that worked into a great example of how to write something that will encourage you to think, reason, visualize, apply everything you know even if it seems way out in left field. Well, here I am going to do the same thing. I think you'll find it informative.

"OK, already, so what the hell is it"? The device I refer to here is called a "Listen down the line amplifier." A simple inexpensive audio amplifier I connected to the phone line through two .01 microfarad capacitors. Worked like a charm—I could hear anything on the line without taking it to an "off-hook" condition. No telltale click. Alexander Graham Poe would have loved it.

OK, back to the story. One day, a few weeks ago, I started to make a call to check my voice mail. I punched the speakerphone button to get the dial tone, and what to my wondering ears should be heard? Someone else checking voice mail. On my phone line. Hmmm. Hmmm, not as in dial tone; hmmm, as in what the hell is going on here? Someone getting into my voice mail? I listen, and hear "Press 7 to delete this message, press star to hear the next unplayed message . . ." "Press 7"? Nope, this is not my mailbox. This was not nonphysical crosstalk—one line "leaking" audio into another. It was too loud and clear, no static. But lines get physically "crossed" sometimes, so I didn't give it much thought.

I had pretty much forgotten the incident when the same thing happened again. I hit the speakerphone button and heard Touch Tones. So, I started thinking I should call repair service when I heard something that I should not have heard. Someone, I still don't know who, had called a seven-digit number which I quickly realized was ANI; a service of the telco. You call the ANI number and the computerized voice reads back the number of the line the call was made from. Now this is weird. It read back my number. The ANI number went in so fast and the digits were so evenly spaced that it had to have come from a memory dialer. Or a lineman's test "butt" set.

Thoughts

Very few people even know what ANI is, let alone have the number(s), and of those, even fewer would have it programmed into a phone with a speed dialer. And unless whoever it is has access to my line through a "sleeper" (about which you will read more later) at his home or office, then the ANI call was made at a B-box or junction point. It wouldn't likely have been from the central office (CO). There would be no reason to use my particular line unless there had been a problem with it. There hadn't been, but now there was.

So I dropped everything and got to work. First, I connected an unused amplified computer speaker to the line, as described above, through two capacitors. I had just gotten it connected when the speaker came alive with a number being dialed, and then an answering machine/voice mail system saying, "You have reached the China (unintelligible) . . . your call will be answered in the order in which it was received." I listened for a few minutes, not hearing any background sounds, and then whoever the caller was decided not to wait any longer and disconnected.

At this point I am very puzzled.

Using another line, I called repair service and got a long voice mail system: "Please select from the following choices . . ." (Pacific Bell loves to offer you choy-suz). The computerized voice ran a check on the line, and after a couple of minutes it proudly announced (as proudly as is possible for a synthesized voice) it had determined that there was a problem on the line. No kidding! I never would have known. An appointment was scheduled.

A few minutes later, another call was being made. I had been frantically digging through boxes of stuff, looking for something, and I'd found it and was in the process of getting it hooked up.

Just in time. The big red LEDs on the DTMF (Touch-Tone) decoder displayed an 800 number. Again, the tones were fast and evenly spaced, meaning they came from a speed/memory dialer. I heard a ring, then an answer . . . an audio tone. It wasn't a fax line, since there were no warbling, handshaking tones. Then I heard the rushing sound of analog data being transmitted but could tell that it was too slow to be an ordinary computer modem. Sounded like 2400 baud. The call lasted from 13:10 to 13:14 (on 14 January 1999) and then disconnected. Who the hell was making calls on my line? Time to investigate.

First, I went to a pay phone and called the 800 number, just to see if it would answer a voice call. Nope. Got the modem tone.

Then I called repair (again) and tried to explain all this, and asked who the 800 number was assigned to. Naturally, they gave me the runaround, shuffling me back and forth, so I called the Pac Bell business office where I was instantly connected to a real live human. More or less. I could tell from the very beginning that I had been instantly relegated to that class of weirdos who claim that their phone is tapped by (1) CIA, (2) martians, (3) gremlins, (4) the metalones, or (5) the recently and dearly departed who are trying to reach them from the "other side." OK, look—I am not one of those people who love to bash Ma Bell. I like the phone company. At least as far as the technology goes. But I don't like dealing with administrative people who would have great difficulty pouring artificially colored lemon Kool-Aid out of a tennis shoe if the instructions were stamped on the sole. Sheesh.

After a long wait, I was advised that while this was, in fact, a Pac Bell 800 number, they would not reveal the name of the subscriber. I was determined to find out where that 800 number was and what it was for, and

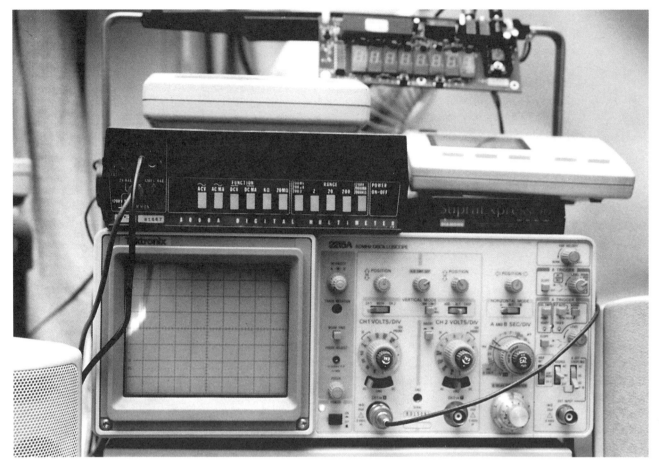

Part of the setup used when I had the little problem with Ma Bell. The Tek 2215 scope on the bottom, Fluke DMM on left, and the DTMF decoder hanging from the scope handle. I didn't have anything to mount it in so I jerry-rigged it using pipe cleaners. On the right is a Diamond 56K modem and Pac Bell Caller ID display.

when I get intensely interested in a bit of information, I usually get it. It was time for a little research and perhaps some social engineering. I got on the Internet and did a lookup, but the 800 number wasn't listed.

Shortly after that, the line went dead. I connected the fluke to the offending line. Zero volts. Should be about 48. Hmmm. I decided to go run a few errands. When I got back, the first thing I noticed was the fluke. The line voltage was back. I hit the speakerphone button and got a dial tone. It was working again.

I called repair service (again) and, after some time on hold, was again connected to another real live human. After assuring the tech who answered that I was not drunk, do not use LSD, and had not recently escaped from the Funny Farm, I was connected to a supervisor. I told them that while it was indeed working, I still wanted someone to come out and check the 66 block. Then I explained why I was so curious about the ANI call. I was told that sometimes the repair people used ANI in their job, and I explained that I know this, but want to know why a tech would ANI my particular line; I hadn't reported trouble before this happened.

I was advised that a repair person would be out the next day.

Dialog with Ma Bell

The next day I was taking a break, reading a John Grisham novel, and I heard Touch-Tones coming from the speaker. I jumped up and looked at the display just as I heard the ANI synthesized voice announcing my number. There it was, the ANI, which is 211-XXXX, in big red LED numbers, courtesy of the DTMF decoder from MoTron Electronics. A minute later, I heard the buzzing sound from the speaker, which was the 90-volt ring signal, and then my phone rang. The conversation went like this:

"Yo."

"Is this ###-####?"

"You already knew that. I heard you ANI me." (I always was a smart-ass.)

Long pause. "How'd you do that?"

I explained all of this to him. Tech's name was Joseph W., I found out later. He said he was at the B-box half a block away and would be out in front in a few minutes. I let him in and he checked the wiring on the 66 block. Everything was normal. There were two possible explanations. The wires that are used for a line may be configured in different ways. In this case, I later learned, there was a 1,200-pair F1 feeder cable that went underground directly from the Steiner CO to the B-box down the street. There were no other appearances—places where the line could be accessed. From the B-box, a distribution cable went directly to my apartment building.

However: my pair didn't necessarily end at the 66 block in the basement. It is possible that it may have been used for a different number at some time in the past and so may have continued on to another 66 block in any number of other buildings. Across the street, down the block, wherever. If so, the pair was not looped at my 66 block; it was cut so that the wires leading from the B-box to my building's 66 block went only to my apartment. At least this is how it was supposed to be. Also, my pair might have gone out on a different distribution cable. To another building. This would be from the B-box, and normally there is a "jumper" that is removed. But no one is perfect, and mistakes are sometimes made. These other locations are called "multiples," but if it were a multiple in this case, then someone, somewhere in another building had access to my line— there was a dial tone there. This is called a "sleeper on the line." Perhaps someone who'd just moved in and hadn't arranged phone service yet just happened to plug the phone in and discover it was working.

Now, another possibility was explained to me. In this neighborhood, there was a great deal of construction going on (ever been serenaded by jackhammers at breakfast?) and a lot of old copper wire cables were being replaced with fiberoptics. So, whenever a pair was switched over from the old to the new, the technicians had to verify that they were making the right connections. And, presumably, they used ANI. Maybe that's why I heard someone ANI my line. . . . But, as I said, I learned about the F1 feeder cable. This new construction was not the cause— it didn't involve the feeder cable.

Meanwhile, the repair guy said he would check for multiples and call back in about an hour, which he did. "There aren't any," he told me, "and even if there were, they would have been disconnected [the jumpers removed] at the B-box." "So the problem was most likely due to this new cable installation." Also, I was advised that I would be contacted by a supervisor later that day or the next.

Now, back to finding out about this 800 number. The sound I heard when someone called it was data being sent at a low baud rate. So, this is too slow to be an Internet connection. I tried it anyway, but I didn't connect to anything. Didn't really expect to but wanted to eliminate one possibility.

Next, I downloaded a BBS program and configured it, and punched in the number. It answered but I got only garbage on the screen—what appeared to be random characters. So I played with the configuration, trying VT-100 and ANSI settings. No good. I changed the 8th bit to off. No good. Garbage. I made a few other changes, and Viola! On the screen I saw: Port 1.71 THIS IS A PRIVATE COMPUTER. UNAUTHORIZED ACCESS WILL BE INVESTIGATED AND PROSECUTED. Login:

I didn't login. I didn't try and hack my way into whatever this was. I quickly disconnected. But I thought I had at least part of the answer.

The following evening, I got a call from a James T., a Pac Bell supervisor. He explained that he was following up on this. I was surprised, and said so. He said that when someone complains often enough, they have to look into the problem more closely, and apparently "I had been raising all kinds of hell." Maybe it was because I told one of the administrative people that I was processing proprietary information that was protected by federal law and, if necessary, would ask my attorney to look into the matter. I didn't say that this "proprietary" information was a book I was writing and the "federal law" is the First Amendment. And I didn't say the book was about wiretapping. Never, ever use that word with telco people. They don't like it.

Well, by now I was very tired of talking to people who don't know what ANI is or what a loop is, and asked if he was a technician.

"No . . ."

Another view of the setup. My everpresent Optoelectronics Xplorer in front and an old Alinco dual band transceiver in the back. The computer in front is an old Toshiba used to keep a record of any calls made from the line. The fan in the background? It keeps the monitors cooler and increases their life considerably.

Shit. Trying to talk telco-tech to administrative phone people is like trying to order from a menu printed in French when you don't speak the language. "*Monsieur*, here is your dead cat, and the lady's baked potato(e) is in the men's room."

"Uh, look, I really need to talk to someone who . . ."

"But I was for 20 years before I became a supervisor . . ."

Ah! OK, we can communicate. I related part of this to him in terms we both understood.

He asked if I was sure that this wasn't crosstalk, and I told him I had a fluke DMM on the line, and that the DTMF decoder would not have read out the digits if it were—it needs a clear, noise-free audio signal. He understood and was convinced. Said he'd look into it.

He wanted to know how I knew all this stuff and I told him I read a lot. And that I was working on a book that will include a detailed description of what was happening. I could sense a change even before he said anything, then he asked about the book. "Will it be available in bookstores?" he wanted to know. "At Barnes & Noble?" His favorite store, he told me.

He wanted to know if I'd ever worked for Pac Bell. No. I told him about experiences as a kid, taking phones apart, being offered a job at GTE when I became old enough. I asked about REMOBS (REMote OBServation—a method of accessing phone lines by telco technicians for testing purposes and also by feds for unlawful surveillance), while explaining about the phone in the picture in *The Phone Book*. He was very open about this. "Yes, we can listen in on anyone we want any time." I asked if that included the feds and he clammed up. Just like "Joe" in *The Phone Book*.

I asked if someone outside the telco had the REMOBS access numbers, could they do the same thing? "Oh, yes." No hesitation. I asked if the telco is uptight about people knowing this. He said something like, "Yes but not much we can do as it is already well publicized." I knew what he was thinking . . .

He said he would "sponsor" me if I wanted to apply at Pac Bell. Thanks, and while I love the phone company, I am not ready to join Corporate America. Lastly, I asked him, again, to look up the 800 number and see who it is assigned to. This may solve the mystery. He layed the phone down, and after a few minutes was back on the line. He seemed a tad nervous, hesitant, which, again, told me something. Then he admitted what I suspected—that the 800-number was for accessing the Pac Bell switch, the telco computer. He explains that this one is in San Ramon, California. The number that is used to access Computer System for Mainframe Operations (COSMOS), and all sorts of other things.

FANTASIES

Suddenly, I was fantasizing: If I could get into that, I could do all the things that the feds blame on "hackers" when the telco software develops a glitch.

- I could re-assign numbers.
- I could route all of Microsoft's calls to Netscape.
- I could arrange for an unnamed person you already read about to get a phone bill that would emulate the national debt!
- I could track down that kid that took my girlfriend (Karen) away from me in the seventh grade and see to it that he never again would have a working telephone line.

Visions of megalomania flooded through my consciousness—the power I would have! I could rule the world!!

Then another picture filtered into my mind. Of the Secret Service kicking down my door and pointing machine guns at me. That sinking feeling. Back to reality. I attempted a little humor.

"Hell, I have social-engineered myself halfway into the switch." He didn't think that was funny at first, then realized I was kidding and started to laugh. Subdued laughter, like from reporters when presidents make their bad puns at a press conference. Polite.

We talked about that. Fat chance. The terminals that repair techs use have access codes programmed in, and these change frequently, he told me. And, there are other barriers (that he didn't want to talk about), as well as even being able to log on at all. Very unlikely. And, as this is an 800 number, they have a record of every attempt to access the switch. Real time ANI.

I told him that the portable units the maintenance people use are supposed to make these changes automatically and that sometimes the techs scratch their PIN on the case. Or so I hear. "So, if I were to obtain one of these goodies, I could then do all sorts of things, right?" He sighed and said yes, but only until it was reported missing.

I mentioned that I still didn't have the answers I wanted and asked what they could do. "Hey," he said, "you're troubleshooting this thing yourself." "Ah, just applying a little logic," I told him, thinking that this is what most people would do, then remembering that this was wrong. Most people would never even know what the hell was happening, or even that anything was happening in the first place.

Meanwhile, it had been a fascinating conversation (I love yacking with telco people), but James has to go. I hung up and connected something else to my line. Starting to get complicated.

The following Saturday (16 January) at 11:36 A.M., someone made a call from my line to 563-XXXX. An answering machine picked it up announcing that the caller had reached 440-XXXX. Call forwarding? I didn't hear that interrupted ring signal indicating that this was happening. This was a personal machine rather than commercial voice mail. The caller disconnected without leaving a message. So far, I had yet to hear anyone actually speaking, other than recordings, which, all things considered, isn't unusual; many people refuse to leave messages on answering machines.

A few minutes later the same WATS (800) number was called and a data stream lasted from 11:39 to 11:53. Interesting that a call answered by a machine preceded the WATS (800) call as it did in the previous incident. I got on the Internet and checked the numbers. Neither were listed. I called voice. On the original call, to 563-XXXX but which went to 440-XXXX, the answering machine recording was a female voice. This time it was male, " . . . away from my office or on another line."

Strange set of coincidences.

I made a call to the public library and had the numbers checked in the Haines reverse "cross" directory: 563-XXXX returned no record. The 440-XXXX number came back to a business located several blocks away. There were several other people listed at that address, which is a small and very expensive apartment building.

I had errands to run during which I went to a pay phone several miles away and called 440-XXXX.

"Good afternoon."

"Gimme the ____".

Guy is very uptight. Suspicious. "Where did you get this number"?

"Oh, some guy in a bar, said ____."

"He . . . this . . . some guy said we . . . this is the ____, he told you that?"

"Not exactly. I was shoulder surfing."

He didn't hesitate for a second, indicating that he knew what I meant.

"WHERE DID YOU GET THIS . . . WHAT BAR . . . WHO WAS THIS GUY . . .?"

It ain't normal to get that uptight about a wrong number. I disconnected. Something was rotten on Nob Hill. Just for the helluvit, I crossed Jones Street and went halfway down the block. Standing behind a bus shelter, I focused my Nikon telephoto lens on the booth, not really expecting anyone to show up. No one did.

So What Happened Already?

When you go to troubleshoot something, you need to consider all possibilities, no matter how unlikely or even ridiculous, and methodically eliminate them. Whatever is left, if anything, must be the truth. Theoretically. Let's look at some possibilities.

1. *The feds have legally tapped my phone*. Not likely.
   - I am not doing anything that could justify a warrant and the considerable expense you read about. The fact that they don't particularly like the books I publish isn't enough.
   - The feds have all my books and (also for other reasons I will not get into here) know damn well I would no way in hell use my home or office phones to talk about anything I didn't want to be intercepted.
   - A legal tap would probably be in the CO, where it would not be necessary to use ANI; the COSMOS computer is right there in the frame room.
   - It is unlikely that the feds would use my line to check their voice mail. Someone might intercept them!

....

There was a story going around some years ago about how the feds were tapping a line, and to test it an agent used the line to call in to their offices. "How's the weather in Washington?" the agent wanted to know. Well, it seems that the subject of the wiretap had something, perhaps an amplifier like mine, on his line and heard this. Apparently he came on and made some crack about the weather, letting the feds know that he knew, and subsequently blowing their operation out of the proverbial water.

....

2. *The feds or local police have installed an unlawful physical tap*. This might be because of a rather "strange" call I made a couple weeks previous, as part of researching this book, to see what, if anything, would happen. Or, perhaps they are pissed off about what I said on a recent (10 January 1999) interview on Wisconsin Public Radio. Unlikely.

3. *They are using SAS.* This is Surveillance Administration System, a system set up so that law enforcement agencies (and some hackers) can tap into a line in real time without having to use a loop extender. A loop extender is a special dedicated pair of wires that goes from the CO to the feds' offices. In order to tap a particular line, the feds have to get the security people at the switch to make the physical connection from the target line to the extender.

   With SAS, this is no longer necessary. The feds access the switch by computer and type in the number to be tapped, and the SAS makes the connection electronically. This system has been set up, apparently, as part of the CALEA; Communications Assistance to Law Enforcement Act.

   Pac Bell won't admit that SAS exists, even though Pac Bell sells CDs containing technical info about SAS. Very expensive disks, incidentally; about $1,200.

   Possible.

   I suspect they have a list of people they like to check in on now and then, and I may be on it. As mentioned somewhere in this book, I worked for a company that made cellular radio interception systems and other spy stuff. Built and tested them and wrote the operator's manuals for them. Maybe they think I am still involved in this. No way. The owner of the company and one associate are in the slammer and another is on the lam, and I have no desire to join them.

4. *The feds hate me because I have accused them of using unlawful surveillance so they decided to use unlawful surveillance to set me up.* Using a portable terminal (which is what was used to connect to the San Ramon switch), they made the calls from my line to create a record. Evidence. They plan to bust me as a hacker to make an example of me. Maybe I'll get to meet Kevin Mitnick. And while one day I hope to, I'd rather it weren't in the slammer. Unlikely.

5. *Someone (industrial-strength spy or whatever) who happens to have a wrench and has figured out that this is all that is needed to open some bridging boxes, and who has a test set with ANI numbers preprogrammed, has decided to tap my line.* Unlikely. To do so he would have to have a loop extender or an unused line that goes to his home, office, listening post, whatever. Without access to COSMOS, or another way to get cable and pair numbers, how would he find my line and the extender line? Otherwise he would have to use ANI to call however many of the 1,200 lines in the B-box. Or stand there at the B-box with the test set waiting to hear something interesting. And even in San Francisco, sooner or later someone would notice and get suspicious.

6. *Someone who lives in this building and can get access to the phone closet wants to use someone's line to make calls.* Perhaps he just moved in and doesn't have a phone installed yet. Unlikely.
   - Again, the matter of having a test set that has ANI programmed into it. Slim chance.
   - During one of these incidents, I have good reason to know that no one was accessing the phone closet in this building.
   - Phone installation in San Francisco can be done in a few days or less.

7. *What with all the stuff I have connected to my lines, I could somehow have accidentally done this myself.* No way. I do not have any ANI numbers programmed into any of my phones. I don't have any numbers programmed into any of my phones. I just don't use speed dialing. And anyway, at the time, I didn't even know any of the seven-digit ANI numbers. The only one I had was 15 digits. Old, but it still works.

8. *There is a "sleeper on the line."* This is a telco term for a number that has been left turned on at one of the multiples where someone's pair goes into a different building, and the people there discover there is a dial tone and decide to use it. Unlikely. Unless the telco technician was lying to me for whatever reason, there are no multiples on my pair. The 600-pair feeder cable goes underground directly from the CO to the bridging box down the street. Then a smaller distribution cable goes directly to my apartment building.

9. *Repair personnel used my line from the bridging box because it was "handy."* When they need to call into the switch in San Ramon, perhaps to get COSMOS information or whatever they need for whatever reason, or to see what is playing at the Bijou Theater, they just use those particular terminals because it is convenient. Because they are just easier to get to. Probable. A week or so ago, I heard the same thing—someone on my line. The first few digits were pulse (rotary) dial, and the rest DTMF. I was out the door like a flash, hustling down to the B-box while being reminded that the streets of San Francisco contain all manner of things—some sharp, some sticky—that are not conducive to running around barefoot. There was the Pac Bell truck with a guy standing beside it, so I went up and started asking him stuff. Some telco people are very talkative, and he was one of them. He opened the box and showed me that there were two contacts, terminals, on the inside of the door, where the lineman's test set was connected. From there, a short wire with clips on the ends was used to connect to the terminals in the B-box. The wire reached only a few of the connections, and my pair happened to be one of them.

LINE IN

This is a block diagram of the equipment I set up when I had the little problem with the telco. It is not necessary to be this elaborate, although it is fun. Remember to use capacitors between the phone line and speaker and that the DTMF decoder goes to the audio from the speaker and not directly to the line, which will probably destroy it.

However, when I asked why some of the digits were pulse dialed, he clammed up. My DTMF decoder doesn't work with rotary dialing.

OK, seriously, the most likely explanation for most of this is (9) that all of this was merely telco technicians on the job. The voice mail calls could have been personal, something that's frowned upon, but I guess it happens. So, without anything else to do, I more or less concluded that was it. Until a week later when again I heard the dial tone coming from the amplifier speaker. A few tones were keyed in but the gain (volume) was too low to register on the DTMF decoder. Telco repair people again?

At five minutes after 3 A.M.?

The same thing happened a few times after that, and also something new. I heard a very loud click from the speaker, the line voltage went to zero for about two seconds, then jumped to 82 volts, then back to the usual 47. And stayed there. Once a day, or every other day, it would do this this, and always after 8 P.M. I called repair again to see what would happen when they ran the test. The voltage dropped to zero for several seconds, then back to 47 with the speaker clicking several times. This repeated, and I heard the ring signal but the phone didn't ring. Still 47. Then the process repeated for a third time, and this time the phone rang. Ring stopped, voltage dropped to 40, then up to 56 for half a second, then back to 47, but not once did the line voltage go to 82.

Strange.

This could possibly be a DATU (digital audio test unit), but unless the telco had it set up to run periodic checks automatically, then why my line? Another week went by, and I heard another click followed by a series of pure (not DTMF) tones—eight or ten of them, over a period of maybe one second. Later, I would learn that this is a "tone slope." The audio tones are used to check for line losses at different audio frequencies.

So, apparently some of the things you will hear if you set up your own listen-down amplifier, are not so sinister as one might suspect. But someone on the line in the wee small hours?

The diagram above shows how everything is connected.

- The line comes in to the CommShare (made by Command Communications in Colorado), which is a system that lets you use a computer, fax, and phone all on the same line. When a call comes in, it is analyzed and routed to the appropriate device.
- Plugged into the CommShare is an RJ-11 modular splitter which takes the incoming line two places. First, directly to the Fluke digital multi-meter for monitoring the line voltage. The Fluke has very high impedance so it does not interfere with the line, does not affect it in any way. Spliced to that wire is a shielded cable going to the amplifier, through a pair of .01 uf capacitors. This blocks the line voltage—typically 47 volts DC (VDC)—to prevent it from seizing the line (from taking it to an off-hook condition) and damaging the amplifier. The amp is one (right) of a pair of stereo speakers made for use with a computer. An old, cheapo pair I had in the junk box.
- Next is the Touch-Tone decoder. It requires an audio input as the tones are audio, and not a direct connection to the line, which would damage or destroy it. I used a TDD-80 from MoTron and took the audio from the headphone jack of the right speaker. Now, as the jack opens the connection to the speaker, I have to do something so I can hear the sound. So I take the cable that normally goes from the right to the left speaker, solder a 3.5 mm connector, and plug that into the Marantz recorder and use the built-in speaker to listen. Should I hear anyone actually talking on the line, I hit the "pause" button on the Marantz and I've got 'em on tape. (But I never have . . .)
- The DTMF decoder has an RS-232 output that can be connected directly to the serial port of a computer. With the software shipped with the decoder, you can keep a permanent record of every call placed from the line—number dialed, date, and time. For this, I used an old Toshiba T-1,000, one of the very first portable computers that included a floppy disk drive (no hard drive) to appear on the market. Fascinating how technology changes so quickly. That Toshiba, which uses an 8086 processor, cost me more than my DEC Pentium notebook.
- Also coming off the decoder is a TexTronix 2215A oscilloscope.

A sophisticated and very effective system, but not all of it is necessary to be able to just hear the phone line audio. For that, only the speakers and a patch cable are required at a cost of about $20. The MoTron decoder sells for about $100. The Marantz recorder is expensive, but any recorder with an external speaker output will do. Any inexpensive digital meter will work, and the scope isn't at all necessary. I just wanted to look at some waveforms during a particular stage of the operation.

Do Try This at Home

Let's go over this again step by step so that you can set up your own listen-down amplifier.

At a computer store, get a pair of inexpensive stereo "multimedia" speakers. One of the speakers, usually the right channel, the one that has the power switch and volume control, will probably have a cable with a 3.5 mm miniature plug that goes to the sound card. Check it before you leave the store, and if no cable is supplied, there will be a jack—either an RCA (phono) of the type used for speakers on a stereo receiver or the 3.5 mm. Buy a cable that has the right connector for the speaker on one end. The other end doesn't matter.

Next, you need a standard telephone modular RJ-11 extension cable, a "splitter," and two .01 microfarad disk ceramic capacitors.

Take the end of the cable that came with the speaker, or the one you purchased, and cut the connector off of one end. Then, clip the phone cable at one end, leaving a few inches of the wire, and splice it to the audio cable, using the capacitors. They have to be used, otherwise the amplifier will seize the line. You want the unshielded phone cable as short as possible to reduce 60-cycle hum from getting into the speaker. It is better if you solder the connections, but some phone wire uses foil-like material wrapped around a few strands of nylon, which doesn't solder very well. Try tightly wrapping some very fine wire around the foil. You can get this from an old 110-volt extension cord. Strip off a few inches of the insulation and separate the strands, wrap them as tightly as possible, and then solder them. Better yet, use Scotchloks or other solderless connectors like twist-on "wire nuts." Be sure you don't short the wires. This could put your phone out of order until repair service resets it.

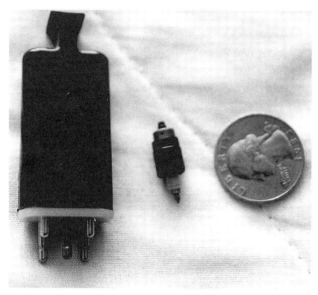

Two types of heat coils (fuses) used in the telco frame room to prevent damage to their equipment. Apparently it isn't unusual for them to "blow," as when I toured the Larkin CO there were hundreds of them lying on little shelves and on the floor under the distribution frame. Some COs have replaced them with circuit breakers that can be reset with the computer.

Plug in the speaker power supply and press the ON switch, making sure the LED comes on, and turn the volume down all the way. Unplug your phone and plug the splitter into the modular wall connector. Plug the phone into one side and the new modular cord that goes to the speaker into the other. Pick up the phone (or push the speakerphone button), adjust the volume on the speaker, and you should hear the dial tone. If not, if you hear a loud hum or buzz or maybe an AM radio station, you have a bad connection, probably an open ground. Recheck the splices. Once it is working, you can set the volume so that you hear the dial tone when the phone is off-hook, but not so loud that you hear the hum or faint background clicks that may be distracting.

Now you are all set up and will hear anything that happens on the line. Including feedback if you forget to turn the volume down when you make a call.

To add the DTMF decoder, remember that it requires an audio signal from the speaker and not a direct connection to the phone line. Use the cable coming from the right speaker (that goes to the left speaker), using whatever type connectors are required. Make sure the volume is turned up slightly (about one third) or the decoder will not have a strong enough signal to work with and the numbers will not display correctly, if at all.

MoTron is the manufacturer of a number of products, including the TDD-8X, which decodes all 16 Touch-Tone digits—0 to 9; # and *; A,B,C, and D—which are used in some military phone systems. It comes assembled and tested, with a large, bright red LED 8-digit display and scroll right and left buttons to view its 104-digit memory. It also has a clear button to erase what is stored. To use it, plug in a 12-volt adapter and a cable from the decoder's audio jack to a scanner or tape recorder. You can also dump its memory to a computer through a serial port to maintain a record of all calls placed from the line it is connected to. MoTron makes excellent equipment and is highly recommended; the surveillance electronics company I used to work for used them exclusively. The display in the photographs is a MoTron.

I just heard from MoTron advising me that the TDD-8X has been replaced with a newer model, the T8 Tone-logger, which is almost the same but uses a different microcontroller and has an amplified speaker output. This simplifies things, making the setup easier.

MoTron Electronics
P.O. Box 2748
Eugene, OR 97402-0280
www.motron.com

PHONE PHREAKING TERMS AND TRICKS

This chapter is based in part on personal knowledge, but mainly on a FAQ written by Sasha Berkman (stukach@yahoo.com), aka Dr. Seuss. I have rewritten it to make it more appropriate for The *Bug Book*, as much of the info in the original document is too detailed and technical. Again, this book is for the "average" person, so experienced phreakers may not learn a great deal here.

So, What Is a Phreaker?

Phreak is short for phone phreak, a hacker of the telephone system. A phreak (or phreaker) is someone who wants to learn about the telephone system. Some people who claim to be phreaks are thieves who do nothing but rip off long-distance service. Others are only interested in sneaky tricks and screwing other people. A phreak is not someone who destroys phone property if it's not necessary for advancing his knowledge of said system. Boxing and other ways of hacking the telco do not make someone a phreak. Being a phreak is a way of life. There are many different views on what is a phreak, so two people may call themselves "phreaks," yet have two totally different viewpoints.

Phreaking and hacking are indeed a way of life—one that has to do with wanting to learn, to understand things, to take them apart and see what makes them tick. Telephone systems, computers, networks, radios, watches, whatever.

Originally, this did not necessarily mean learning about things that we are not supposed to know because there weren't that many things we weren't supposed to know. And people who were able to figure these things out were sometimes offered jobs where they could continue the learning process while contributing to the development of new technology.

But, in the last 20 years or so, this has changed radically. Personal computers have been available to the general public, and telephone systems have gone to electronic switching—replacing the old Strowger switches and crossbars. (See *The Phone Book* for details and photographs.)

Now, in the face of the "New Millennium," more and more things are becoming verboten. Hack your way into much of anything and the only job you are likely to be offered is cleaning toilets in a federal "correctional facility."

In spite of what federal agents, supported by a government-controlled media, would have you believe, phreakers and hackers are not inherently destructive. Matter of fact, few of them are. Those of us who just want to learn more about the system have no use for those who attempt to damage equipment or shut down systems. These are not phreakers; they are criminals.

Personally, I think the telco is the greatest thing since sliced bread, and even though Pac Bell et al are part of Corporate America (for which I otherwise have little love), I have no sympathy for those who try to deliberately destroy it. Hack your way in: yes. Observe: yes. Learn: yes. Steal: no. Destroy: no.

Government and law enforcement agencies cannot understand this, and in the infinite myopia typical of bureaucracies, they manage to turn everything into a lose-lose situation.

Meanwhile, an irresponsible media continues to publish inflammatory stories about "dangerous hackers" that are total fabrications—to describe, in detail, events that never even happened. Like the "hackers" who took control of a military communications satellite and held the government "hostage," the "teenager" who hacked into the computers of several large corporations and held them at his mercy, forcing them to buy him, among other things, a new Ferrari. Bullshit. These things did not happen. And the media will continue to publish such lies, as that is their nature.

Journalistic integrity is becoming an oxymoron, but nontechnical people, not knowing any better, actually believe what the media tells them. Like the lady in an elevator a few weeks ago: "I heard on the news that hackers are going to shut down the entire Internet at midnight."

Good grief.

Where Does a Newbie Start?

"How can I learn about phone phreaking?" By asking this question, you are starting off on the wrong foot; this qualifies you as a lamer.

Read this chapter carefully. Memorize some of the terms.

Read everything else you can find. Books, newsletters, technical manuals.

If you have Internet access, get on the Web and download some of the software and learn to use it. Visit some of the Usenet newsgroups and read the posted messages. Buried under the bullshit and flaming and endless posts asking, "Will someone tell me how to phreak," there is some useful information . . . now and then . . .

Is Phreaking, or Anything Related to It, Legal?

Complicated question, depending on what you do. The feds are coming down very hard on anyone who does much of anything anymore. Toll fraud is a federal offense. If you're caught ripping off phone service ("red boxing," for example), you'll probably be prosecuted.

Accessing a switch (if you can get in) is an extreme no-no, and getting caught will get you extremely busted. Even if all you do is look around, you will be in deep shit. And for most people, looking around is all they will be able to do, since the programming they are confronted with requires some heavy knowledge to understand.

Accessing telco property, such as bridging boxes and junction points, will also probably get you busted if you are caught. Can you say, "I lost my Pac Bell ID card, Officer," and, "Pac Bell truck? Uh, gee, somebody musta stolen it, Officer."

Now, back to getting started. You need to learn some things about telephone companies, starting at the CO.

How Do I Find My Local Central Office? Any Central Office?

There is a program called CO Finder for Windows at www.stuffsoftware.com/cofinder.html. It has an exhaustive list of telco CO locations and information about them. The screen is split into two parts. On the left, you can enter the state, LATA (local area), or NPA and the program will compile a list of all COs in that area along with which features are available at each, the latitude and longitude, and many other tidbits of info. Click on a location on the right side of the screen and the program calculates the distance between the two. Fascinating.

Any selection criteria can be used to help you find what you're looking for. For example, let's say you were trying to find a Frame Relay Packet Switch in Florida close to a location served from Area Code 407, Prefix 292. You would press the NXX button and enter 407 292 on one side of CO Finder. Then you would select FL from the state list and Frame Relay Packet Switch (TD) from the feature list on the other side of CO Finder.

The distance listing shows ORLANDO as being the closest at 6 miles. When you click on that location, all pertinent data is displayed for that switching center, including the CLLI code (ORLDFLMABB0), which is used to order service from the telephone company (BellSouth).

A list of more than 150 standard features is used to designate the options available at each switching center. These are listed as official code designations, but CO Finder lists the English equivalent of the code next to each one. This way the rest of us can use the information, which in the past was only useful to telephone company engineers and other guru types.

The price of CO Finder is $100 and includes the latest available database. For $240 you can get quarterly updates for one year, and for $495, monthly updates for one year. Or, you can download the demo, which is fully functional but has an older database.

If you do not have Internet access you can order by mail or phone:

Stuff Software
3879 Brantley Place Circle
Apopka, FL 32703
www.stuffsoftware.com/cofinder.html

What Is the Inside/Outside Cable Plant?

The inside cable plant refers to all hardware inside the dial tone office, the telco facility that provides you with telephone service, i.e., a dial tone so that you can make and receive calls. The outside cable plant is all cables, wires, bridging boxes, tandem offices, junction points, and other transmission hardware between your phone and the CO.

What Is the Layout of the Cable Plant?

This varies from office to office, but it generally follows a layout similar to this one:

- switch (computer)
- main distribution frame
- concentrator (multiplexer) group
- battery room
- cable vault
- F1 feeder cables
- junction points and bridging boxes
- F2 and other distribution cables
- line protector and demarcation point
- customer premises equipment

What Is a Switch?

The actual telco mainframe computer. There are different types of switches, including

- the dial tone switch, also called the end office
- the toll switch
- the remote switch

Dial tone switches interface directly with your telephone. They provide your dial tone so you can make calls.

Toll switches connect end offices with toll switches, and toll switches with other toll switches. Phone calls typically follow a path like this: when you make a call, the end office will first look to see if it can complete the call internally—if the destination is within its area. If not, it hands off the call to either another end office or a toll switch, depending how far away the number being called is. From the toll office it can proceed to another toll office or an end office for completion.

A remote switch is a large PBX switch slaved to (controlled by) a CO that is a good distance away. The switches are implemented in areas too small to warrant their own offices, but still require their own switch. Remote switches are switches only and carry none of the other computer equipment necessary for a full-scale office.

What Is a Loop?

A pair of wires that goes from one place to another. The wires from the CO to your phone are a loop.

What Is AMA?

Automated message accounting. It and variations thereof CAMA (centralized) and LAMA (local) are part of a system called "caller log," which records all outbound calls from a particular number (with the option to log incoming calls as well) with a time stamp, duration of the call, and type (voice or data). This supposedly also includes calls that don't answer (DA) or are busy (BY).

What Is POTS?

Plain old telephone service; an ordinary residential telephone line.

What Is a PBX?

A PBX (private branch exchange) is a private phone system used by large companies and other institutions that require a flexible internal phone system, such as college campuses. A subset of PBXs are key systems—PBXs with less than 50 users. PBXs consist of a small phone switch (say a DMS 10), a group of outbound trunks that are nothing more than phone lines to the outside (often fractional Ts or even T-1s on the larger systems), and a number of telephones.

What Is a DISA Port and What Is It for?

A DISA (direct inward system access) port is a maintenance feature on a PBX. When you connect to it and input a password, you seize (control) an outbound trunk of that PBX. Hacking DISA ports is a relatively simple and effective way to get free service, plus someone else's number on the ANI controller. Once you find an open port it may be possible to make random guesses at finding a password on some older systems. Most, however, will disconnect you if you make three incorrect attempts.

What Is SS7?

SS7 (Signaling System 7) is a system for telephone offices to communicate with each other. In the good old days, offices would send information about a call's routing by inband signaling (audible tones sent along with your voice). Inband signaling was slow, unreliable, and subject to wild amounts of fraud. Its replacement, SS7, establishes two connections for each call. The first connection is the voice path; this is the connection that carries your voice. The second connection is the signaling path—out-band, the signaling path that transmits digital routing information, billing info, ANI, etc.

Many years ago, it was discovered that it was possible to use a "blue box" on inband signaling lines to make free calls. The procedure (simplified) was to call an 800 number, such as a car rental company, and wait for the person who answers to disconnect. Then a 2,600-cycle tone was injected into the line, which opened it up to make calls.

What Is a Trunk?

A trunk is a fixed line between two telephone offices—a telephone office and a PBX or two PBXs. Trunks may be large fiberoptic lines, and PBX trunks are usually T-1 or fractional T-1 lines.

What Is an Extender?

Among other things, it's a dial-up number used to access certain long-distance services. It's mostly obsolete now, but at one time a person using certain carriers had to call this number, get another dial tone, and then make the LD call. 1010321 is an extender.

What Is an Internal Office?

An office that the general public doesn't know about. Internal offices are usually used to access complex test systems (such as switching control) or in applications where automation would be impractical (such as customer name and address [CNA] offices).

What Is a LATA?

LATAs are the geographical areas where a single RBOC (Remote Bell Operating Company) can connect a call. If a call passes across the boundaries of a LATA it must be handed off to an inter-exchange carrier and then back to another local exchange carrier for completion.

What Are Inter-LATA Carriers?

Inter-LATA is another name for a long distance company, such as AT&T, Sprint, or MCI.

Where Can I Get a List of Inter-LATA Carriers?

Try www.nanpa.com.

What Is NANPA?

North American Numbering Plan Administration.

What Is Remote Observation?

Every CO has test trunks that are used for loop testing. Now, almost all test trunks do a "busy line verification" (BLV) to see if a particular line is being used before allowing tests to be run, in order to avoid interrupting a conversation. Otherwise, you might suddenly hear a very loud 1004 Hz tone injected into your line. If you build the listen-down amplifier, you may hear the faint clicks as the line is checked for activity, and you may also hear that blasted tone.

Dr. Seuss: There is a certain group of test trunks in every CO that don't check for line activity before they engage; these are called "NO TEST" trunks.

If you "drop" a NO TEST trunk onto an active pair you'll have a seamless (and to the best of my knowledge, undetectable) wiretap for as long as you would want. This is how operators break into conversations with emergency interrupts. I read in an old *2600* that with access to a switch maintenance channel you could add a third trunk onto a line, tap that trunk (don't ask me how; it was a LONG time ago) and have a perfect wiretap.

There's another way: the Dracon Harris DATU. The device is placed in the CO, on the distribution frame, and is accessed from outside by telco technicians, or phreakers who know how. It has its own dial-up number (the number you call to connect to it) and a personal ID number (PIN) is required to access it once the connection is made.

Once you are in, you are dropped into a "channel" from where you can switch to other channels and do various and sundry things. Call a number that is served by that CO, and then you can cause the line to open and short to make resistance measurements and other tests, and also use the line monitor option to hear any conversation on that line. Right, a remote wiretap; REMOBS.

Now there is apparently a frequency inversion (FI) scrambler built into the DATU, so you can tell that there is a conversation going on but cannot understand any of it. However, if this is plain old fixed basepoint, it can be easily converted back to the clear mode by playing the audio into another FI scrambler. If it uses a "nonstandard" base frequency, the audio can possibly be recorded to disk and decrypted with a program available on the 'net. The filename is invert1.exe, and it was once available on the alt.radio.scanner newsgroup, but I haven't seen it there for several months.

If it uses a constantly changing base frequency, such as the devices made by TransCrypt International in Nebraska, forget it. Unless you have intimate knowledge of how they work and a very fast work station or mainframe at your disposal.

To learn more about the DATU, check out www.commprod.harris.com/test-mgmt/lms/.

What Is COSMOS?

Computer System for Mainframe Operations, a database that maintains records of customer accounts such as line cable and pair numbers, long distance carrier used, street address, the cable path (route) and appearances such as B-boxes, junction points, etc. Apparently, if one can access COSMOS, it's possible to make changes, like reassigning numbers. This is what the Legion of Doom is supposed to have used when they allegedly got into the Bell South system some years ago.

What Is a Red Box?

A device that generates the tone that pay phones use to tell the system you have inserted a quarter. They can be made from a "pocket dialer" that stores the Touch Tones for numbers you call often. It has a small transducer (speaker) and to make a call, once you have the dial tone, hold the device to the mouthpiece and press the appropriate button. Once modified, the device plays these tones into the phone, fooling it into thinking you have deposited money.

Do They Really Work?

Apparently they do on some phones, mainly Fortresses, or Bell System pay phones. But not on COCOTs (see below).

Is This Legal?

Hell no. Using a "red box" probably constitutes possession of a counterfeiting device (18 USC 2512) and is definitely theft of products and services, toll fraud, and whatever else. Get caught and you do not get to collect

$200 on the way to a federal lockup where you may spend a great deal of time. And even though the telcos probably exaggerate considerably the losses due to "red boxing," this is still stealing.

What Is a COCOT?

COCOT is an acronym for customer owned coin operated telephone. This is a phone that is privately owned. COCOTs are known for their high security and poor reliability. For instance, you cannot "red box" off a COCOT. They are also known for their high rates. Calls from a COCOT are often more expensive and some even charge for 411.

What Is Coin Signaling?

A signal, sent from a pay phone to the CO, that says enough coins have been deposited to pay for the call.

What Is Ground Detect?

A system that physically senses that a coin has been deposited. It supposedly defeats "red boxing," at least for the first coin. After that the box is supposed to work. This is theoretical; I don't know if it really even exists.

What Is Scanning?

The use of a computer program (sometimes called a "wargames dialer" or "wardialer" from that technically inaccurate movie) to call a series of phone numbers, looking for something such as modem or fax tones, telco test numbers, or long-distance access extenders.

What Are These Programs?

Where can I find them? How well do they work? There are, or at least used to be, many of them available such as ToneLoc (tone locator). If you have Internet access, a search should turn up any number of them, but they come and go. As to how well they work, I have no current info on this. I tried out several programs a few years back and they worked fine. When the program found a modem/fax tone, it made a record of the number, date, and time.

Does the Telco Have a Way of Catching Me If I Use a Wardialer?

There is, allegedly, a system called Overlord that alerts telco personnel when a certain number of sequential numbers, or an unusual number of calls, are dialed from a particular line. I don't know if this is true. I also do not know if scanning is unlawful.

How Do I Avoid Getting Caught?

There are programs that dial the numbers in random order and at random intervals. They may or may not defeat Overlord. No matter what program you use, the more you use it, the more likely you are to be noticed by the telco, and the less you use it, the better your chances of being able to continue scanning.

What Is Trashing?

Another name for Dumpster diving, trashing is the practice of digging through people's trash for credit card information, damaging personal information, useful goods that have been thrown out carelessly, for the fun of it, etc. In the phreaking sense, trashing is done to gather telco documents, phone numbers, equipment, and the always treasured Bell hard hat. Some phreaks also trash other places such as electronics stores to try and find equipment. The most popular place phreaks trash is the CO. Trashing can be fun, especially when you make a nice find.

Is Trashing Illegal?

Maybe. Recent federal court decisions state that once a person throws something away, they have no right to prevent anyone from looking through it or taking it. This is, of course, so that law enforcement can use it as evidence. Some states have allegedly passed laws against Dumpster diving, and there are laws against trespassing. So, unless you know whether or not such laws apply, you take your chances.

The more harmless you look, the more likely you are to be ignored. Who pays any attention to a street person looking for bottles and cans? And in any case, at worst the street person is just going to be told to leave. If you get caught trashing, be polite and explain that you're just looking for a cardboard box to put stuff in. But I don't recommend this within 500 feet of a federal courthouse.

What Is ANI?

Automatic Number Identification. It is a service feature in which the number of a calling station (telephone) is automatically obtained.

What Is ANI II?

ANI II is an additional feature of ANI. It adds a pair of digits to the ANI readout that labels what type of service the number is (i.e., if it's a pay phone, a PBX line, etc.). A complete list of ANI II digits (00 to 99) can be obtained at www.nanpa.com. Here are a few examples:

- 00—Plain Old Telephone Service. Noncoin service requiring no special treatment.
- 02—ANI failure. The originating switching system indicates (by the "02" code) to the receiving office that the calling station has not been identified. If the receiving switching system routes the call to a CAMA or Operator Services System, the calling number may be verbally obtained and manually recorded. If manual operator identification is not available, the receiving switching system (e.g., an inter-LATA carrier without operator capabilities) may reject the call.
- 12-19—Not assignable. Conflict with international outpulsing code. Interesting. The code I got on my line was 18.
- 29—Prison/inmate service. The ANI II digit pair 29 is used to designate lines within a confinement/detention facility that are intended for inmate/detainee use and require outward call screening and restriction (e.g., 0+ collect-only service). A confinement/detention facility may be defined as including, but not limited to, federal, state and/or local prisons, juvenile facilities, immigration and naturalization confinement/detention facilities, etc., which are under the administration of federal, state, city, county, or other governmental agencies.

If you find an ANI II number that works you will hear something like:

ARU ID is Echo Six Romeo
Line Number is 18
Call Interactive Number is 9800
ANI is 415-XXX-XXXX
(the number you are calling from)

What Is Real-Time ANI?

A system, usually on 800 and 888 and 900 numbers, where the number of the line calling them is instantly displayed on a computer screen or other device.

What's a "Dark Call"?

A call to a real-time ANI subscriber where the number ID fails to work. A Dark Call triggers ONI.

What Is ONI?

Operator number identification; that is, when a live operator asks you for the phone number you're calling from.

What Is ANAC?

ANAC stands for Automatic Number Announcement Circuit and is the system that tells you the number you are calling from. This has a variety of uses. Linemen call it to find out the number of the line they are working on. Wiretappers use it to make sure they have the right line. I use it as a reality check. I call it now and then, and if I get the right numbers I know writing this book hasn't totally befuddled my mind.

What Are Some ANAC Numbers?

These numbers are proprietary and cannot be published here. Also, they change periodically, so they wouldn't be working by the time this book is released. ANAC numbers are both local, frequently using prefixes such as 211, 221, etc., and nationwide, usually with 800 or 888 prefixes. Remember that 800/888 numbers may have real-time ANI. So, as you are ANIing yourself, so is Ma Bell.

What Are Test Numbers?

Numbers that access testing equipment or test features set up by the phone company. A number that accesses the Harris DATU, for example.

What Are Some Common
Test Numbers and Their Uses?

Test numbers are assigned to all (as far as I know) of the prefixes and start or end with the digits 00. For example, 415-923-0000 to 923-0099. Test numbers are not limited to starting with 00, and no doubt there are others since 00 is so well known. You can experiment and see what you find, and while I don't know this is specifically prohibited by the telco, if you are discovered they may advise you to cease and desist. Some things you may find are:

- Sweep tones ranging from 304 Hz to 3204 Hz. A common use for sweep tones is allegedly to check for infinity-transmitter taps.
- Milliwatt test. A 1004 Hz tone sent out at 0 dB. Milliwatt tests are used to check for line loss.
- Quiet termination. This feature connects the caller to a port (a "door" into the telco computer) with fixed resistance, 600 ohms or 900 ohms being the most common. There should be nothing but dead silence on connection. Clicks, static, or crosstalk will be clearly evident if a noisy line is used to dial this test.
- Ringback. This calls back the originating number; when you call it and hang up, it calls you back. Then, with some numbers, you can punch in all the Touch-Tone digits in order and the test number will generate two tones telling you that the keypad is working OK.
- Loops. A pair of numbers separated by one digit, e.g., 221-0034 and 221-0035. Call one and you'll get a tone. Call the other and you get dead silence. If both are called at the same time they connect together. It used to be that you could then talk over this connection, but now there are filters that block speech placed on most loops.
- Remote sensors. You may stumble across some sort of remote monitoring system where you will hear a computer-generated voice that announces the time, temperature in the monitored area, status of various equipment, etc. Kinda interesting but not really useful for anything.

What Is CNA?

A service of the telco that provides customer names and addresses as well as phone numbers. Provide a name or address and get the number. Provide the number and get the name and address.

All telcos have CNA, which has all numbers (listed and unlisted) and is available to telco personnel only. Unless you can social engineer your way into it, that is.

Some telcos have a CNA that is available to the public, but only for listed numbers. A few, as of this writing, are Unidirectory at 900-933-3330 and Telename at 900-884-1212. Both cost a buck a minute.

If you are in area code (AC) 312 or 708, Ameritech has a pay-for-play CNA service available to the general public. The number is 796-9600. The cost is $.35 per call, for which you can look up two numbers.

In the 415 AC, Pacific Bell will offer a public access CNA service. It was supposed to be working in August of 1999, but as of this writing (October 1999) it wasn't yet. The dial-up is 415-705-9299. Naturally, this will not include unlisted numbers.

If you are in Bell Atlantic territory, you can call 201-555-5454 or 908-555-5454 for automated CNA information. The cost is $.50 per call, and you can get info on three numbers.

If you know the telephone number you can also do reverse lookups using Database America at www.databaseamerica.com/html/gpfind.htm.

Area Codes

Who assigns ACs?

Bellcore used to issue ACs. Now Lockheed Martin does, but it's the FCC that has final say in any telecom-related matter. What we've lost in the way of tradition we gained in accessibility. Lockheed Martin is very open with its info, while Bellcore insists on charging ridiculous amounts for its paperwork. All their public documents are on nanpa.com

What are the special ACs and what are they for?

- 200: rumored to be reserved for test purposes
- 300: rumored to be reserved for test purposes
- 400: rumored to be reserved for test purposes
- 456: international inbound routing
- 500: follow 'em forwarding services
- 600: ISDN (Integrated Services Digital Network)
- 700: carrier defined (all sorts of fun and games here)
- 710: U.S. Government (only two numbers in the entire area code!)
- 800/888/877: toll-free services
- 866/855: reserved for future toll-free services
- **900: You already know what these are for!**

What Are Test Prefixes?

Test prefixes are exchanges (the first three digits of the phone number) reserved for special purposes such as testing, special routing, teletype (TTY) access, etc.

- 555 is reserved for special purposes such as directory assistance, pay-for-play CNA, etc.
- 959 is a holdout from the Ma Bell days and is supposedly still reserved for test purposes. We've had some bizarre findings here.
- 855 is reserved for TTY services.

Are There Unpublished (Secret) Exchanges?

Yes, there are exchanges that are not published but are still in use for various purposes. Some sensitive test numbers are likely in hidden exchanges. Perhaps more remote observation?

How Do I Find Unpublished (Secret) Exchanges?

You might find this, and other fascinating info in telco Dumpsters. A better way to fetch special exchanges is to go to nanpa.com and download the CO code assignments in whatever area, and compare the utilized exchange list against a list of published exchanges. Find out what isn't listed. Presently, only California and Nevada exchanges are available.

Uh, OK, Now What?

You know some terms, some definitions, something to get you started. Get on the Internet. Check out some of the sites listed below. Read messages. Read more messages. Compare what you read with what you already know from having read this book. Attend 2600 meetings held on the first Friday of the month in many cities. Locations are at www.2600.com. Observe. Listen. Get to know people.

Then start asking questions.

SOURCES OF INFORMATION

Usenet

Usenet dates back to the early years of the Internet, when it was not accessible by the general public. It is a collection of discussion areas called newsgroups on different subjects where people can share information. Today, Usenet is often called the Looney Bin of the Internet, and for good reason. Much of the information you find on most of the thousands of newsgroups is garbage. However, as stated above, if you sift through the bullshit you will now and then find something useful:

- Alt.phreaking
- Alt.hackers
- alt.hack.nl
- alt.hacker
- de.org.ccc—German H/P newsgroup run mainly by the Chaos Computer Club

Any newsgroup that has 2600 as part of the name is a zoo. A waste of time. Next to nothing useful will be found there.

Web Sites

Linenoise
http://www.linenoise.org/
Purgatory
http://soli.inav.net/~dustinm/

The Library
http://www.sonic.net/~theruler/txt/index.html

Gold Matrix's Webpage
http://pages.prodigy.com/Zachs

Jade Dragon's Phreaking Page
http://free.prohosting.com/~jadedrgn/

Digital Misfit's Syndicate
http://roo.unixnet.org/~dms/

Canadian Phreak House
http://mypage.direct.ca/z/zepka/main.htm

Dake Zone
http://www.omedia.ch/pages/dake/idxang.htm

Chaos Computer Club
http://berlin.ccc.de/

Phone Losers of America
http://www.phonelosers.org

Ocean County Phone Punx (OCPP)
http://ocpp.home.ml.org
This is the "home" of Dr. Seuss.

System Failure
http://www.sysfail.org

Phrack
http://www.phrack.com
Some excellent info here, although most of it is ancient.

Government and Business Sites
The FCC
the government agency that regulates us. Take a peek at its site, as it publishes some neat stuff.
http://www.fcc.gov

Telcordia Technologies, formerly Bellcore Search
Info on very theoretical stuff, mostly switching.
http://www.bellcore.com/BC.dynjava?PowerNavigatio nAndSearchPNASGeneralPowerNavigationAndSearch

Telecom Archives
http://hyperarchive.lcs.mit.edu/telecom-archives/
This page is an archive of the comp.dcom.telecom newsgroup. The FAQ is excellent, the articles are good, and if all else fails you can post to the newsgroup.

Telecom Information Resources

http://www.spp.umich.edu/telecom/technical-info.html

This is an enormous site with tons of information about anything to do with phones, but it is not actually a phreaking site.

PacBell Search

http://www.pacbell.com/ir/search/index.html

Surprisingly helpful, PacBell search will outline lots of inter-LATA carrier information for you (including the law), COCOTs, and other sundry phone-related info.

LexiCat Search Demo

This site is a real gem. It offers a searchable index of terms (it cross references everything), as well as articles and reports on related topics. Warning: This is a demo for a product. After 10 searches it resets itself and won't allow you back. Reload the page after every few searches or else get cut off.

http://www.tra.com/cgi-bin/ft-LexiMot/ID= 19970912152925603/lexi7800.html

Blackbox Search

Try this search if you need info on LANs or direct connection. This is an online catalog, but you can still extract enough useful stuff to make going here worthwhile.

http://www.blackbox.com

Lockheed Martin

http://www.nanpa.com/

Country/Area/City/Code/Decoder

http://www.xmission.com/~americom/aclookup.html

Pretty self-explanatory.

AT&T Toll Free Directory

http://att.net/dir800/

A free service to look up numbers. One of the better ones.

Database America

http://www.databaseamerica.com/html/gpfind.htm

Another Web-based reverse directory.

Periodicals

Root Zine
P.O. Box 1178
Maplewood, NJ 07040
http://pantera.openix.com/~mutter/

Blacklisted 411
Subscription Dept.
P.O. Box 2506
Cypress, CA 90630
Published quarterly. Yearly subscription: US $20, Canada $24, Foreign $35. Pay by credit card, check, or money order

2600 Magazine: The Hacker Quarterly
2600 comes out four times a year.
Subscriptions are US $21 for a year in the United States and Canada; $30 elsewhere on the planet. A lifetime subscription costs US $260 and entitles you to every future issue, two T-shirts, and back issues from 1984 to 1986.

Check or money order; no credit cards please.

Subscriptions
2600 Magazine
P.O. Box 752
Middle Island, NY 11953

12
Countermeasures

HAS THEE BUGGED THYSELF?

Before we look into the possibility of thy neighbor's bugging thee, consider that thee may have bugged thyself. Let's start with the electronic baby monitors mentioned above. Although the range is supposed to be limited to about a hundred feet, you now know better than that. And since people are becoming aware of this, they may not use them. Unless someone gives them a false sense of security.

The Strange Case of the Bugged Babies

While making my rounds of pawn shops and second-hand stores, I saw a Fisher-Price "Direct Link Privacy Monitor" at Goodwill. It appeared to be unused, but on close inspection I could see that the box had been opened and resealed. The price was $15.99, so I bought it to see just how much privacy it really afforded. What I discovered might be what the original owner found out, and why it was donated to a thrift shop.

There are switches on the front of both the transmitter and receiver, marked "A" and "B" for the two channels. Aha! Cordless phone channels, I suspect. I plugged the transmitter in, turned it on, and watched the display on the Optoelectronics Xplorer. 49.86 MHz. Yep, cordless phone frequency. Then it started squealing. Feedback. Plain old ordinary audio. I punched 49.86 into the PRO-2006 scanner and moved the transmitter around. Strong signal, excellent audio quality, but privacy? Not. This product is like the "high security" telephone scramblers that were sold by "spy" shops a few years ago. They claim that they can not be intercepted because of the 50-odd thousand scrambling codes they can use. This baby monitor offers 31 codes.

The owner's manual is honest enough to state that this will "minimize the risk of unintentional eavesdropping." But this means nothing to spies, who are very intentional. The signal can be received on any scanner or communications receiver. Unscrambled. Plain, clear speech.

So, what about these "31 codes"? This means that the signal will be scrambled only if received by another Fisher-Price system that is not using the same code. Technically, and perhaps legally, Fisher-Price covers its ass by using the word unintentional. But this is misleading; nothing but a false sense of security.

Do you have a "wireless intercom" or extension telephones that plug into the power line? These are "subcarrier current" devices, and while some manufacturers claim they are private and secure, they are not. See *The Phone Book* for details.

If you live in an apartment building that has a telephone system at the front door, it is possible for someone to find out your phone number by recording the DTMF tones generated when the "code" listed for your name is entered. Then the recording can be played into a DTMF decoder. If your system generates these tones and plays them through the intercom speaker, raise some hell with the management and force them to go to a secure system.

The Bug in Your Pocket

There is another way that you have probably bugged yourself. No, not with a microphone and transmitter that can intercept your dinner table conversations, but something that could reveal a great deal about your personal and business affairs. It's small, battery-powered, clips to your belt, and you cannot escape from it. Right, your pager.

In intelligence, it is often useful to know who is talking to whom, even though it may not be possible to intercept the actual conversation.

Is it legal to intercept people's commercial pager messages? No.

"Yeah, well," you say, "people should have the right to secure pager messages. They shouldn't have to be concerned that 'hackers can intercept them." Agreed. Pager messages should be secure against electronic eavesdropping. So should analog cellular telephone conversations. But they are not. And even PCS (Personal Communications System), with the "Smart Chip," isn't as secure as it is advertised to be. The problem here is that the vendors are not concerned with your privacy. If they were, the systems would have been designed and implemented so that you, the end user, would have privacy and security. But they did not do this because it would have been a little more expensive. It would have cut into their profits. To Corporate America, profit is important.

You are not.

Intercepting pager messages can be a simple operation, and it can be rather sophisticated. It can be dirt cheap, and it can cost thousands of dollars. Federal agents, for example, use very expensive and sophisticated machines that can monitor a number of frequencies at once. And getting your number, capcode, and frequency is a simple matter of asking the vendors. But this is not about the feds, or industrial espionage specialists; it is about how the hobbyist or amateur spy can monitor your messages . . . and, over a period of time, compile a detailed dossier on you.

The Decoder

Let's start with the mechanics of intercepting the pager signal. It is possible to build any of several decoders that convert the radio transmission into text on a computer screen. The cost is about $10 in parts. One type needs only a single integrated circuit (IC), four diodes, and a few resistors and capacitors. Another, the "4 level," uses two chips. Both require some knowledge of electronics to build, but only at the hobbyist level. This is not so complex that engineering or technician-level experience is required. There are also commercially made decoders available (for now) from several sources, which run from $25 to $160 and which accomplish the same thing as the homemade units. They will, with the appropriate software, decode with varying degrees of success the following:

Pocsag 512, 1024, and 2048
Golay
Flex A,B,C, and D
ReFlex
Ardis
Mobitex

The last two are actually data terminals rather than pagers. The others cover about 95 percent of all paging formats.

Incidentally, these decoders have other uses. For one, the ACARS system, which can track the position of commercial aircraft. With some software, you can have a map on the monitor screen showing the location and other information, as well as listen to the pilot's transmissions. Fascinating. Far as I know, this is legal to do, at least for now. Another use is for paging on amateur radio frequencies. Such transmissions are exempt from the Communications Act of 1934 and other laws that restrict what people may or may not listen to.

The Software

There used to be a number of programs that displayed pager messages on a computer screen. The early ones, three years or so ago, were DOS only, no Windows. They were somewhat crude, but they worked. Then, perhaps a year or so ago, someone came out with a Windows 95 application and another program, produced in Germany, was released that was intended for use on ham radio channels.

However, these programs have disappeared in the last month or so. It appears that one program that decoded Motorola's Flex pagers allegedly used proprietary algorithms belonging to Motorola. The author found them, the information needed to produce the program, on Motorola's Web site and apparently wasn't aware that he was not allowed to use it. After having been contacted by Motorola and asked to, he has voluntarily withdrawn the program. As far as I know, it is not available anywhere on the Internet at this time. However, I suspect that someone, somewhere, will provide these programs in the future.

Some of these programs do not require a decoder; the radio can be connected directly to the computer through the Multimedia sound card. Sound Blaster, etc. (However, it is necessary to modify the radio so that the operator can tap into the "baseband" audio, or the "discriminator" output. As with building the decoders, this requires some knowledge of electronics to accomplish.

The Radio

Virtually any receiver that tunes pager frequencies will work but will require some minor electronic modification as described above—tapping into the discriminator or baseband output.

Put It All Together

"OK, someone has the equipment and wants to target me. How do they go about this?"

Again, this may be a very simple operation, or it may be very sophisticated. I will explain: Monitoring pager messages at random—whatever happens to appear on the monitor screen—is simple. Monitoring the pager of a specific person is not so easy, but it can be done.

All pagers have an electronic serial number, called a capcode, which is used to identify them, the same as the ESN identifies cellular telephones. Capcodes are engraved or stamped on an adhesive label attached to the pager and also, sometimes, the name of the vendor—Airtouch, Sky Page, etc.

If someone examines your pager, they can copy down this information, and so, having your capcode, they have a start toward monitoring you. Some pager intercept software has the ability to search through the messages received and display—on the screen or in a separate window on the monitor— only messages sent to your (any) capcode. They can be viewed as they are received and stored on disk for later review.

Capcode Only

With only the capcode, it is possible to scan all pager channels until it pops up. However, this is very time consuming, depending on how often someone beeps you. If you get dozens of pages a day, this could be done in a day. If you are beeped only once or twice a week, it could take months, and someone who is determined to monitor you isn't gonna wait that long.

The next logical step would be to determine, from your capcode, the frequency used. I do not know that there is a database of such information available, except to employees of the vendors, so I don't know how this can be done. Other than having a source of inside information.

Vendor

If they have the vendor's name, they can scan the many pager frequencies until they find the right one, and then plug in your capcode and set the software to flag it.

Ever wonder how pagers vibrate without making a lot of racket? Inside is a little motor, similar to the ones used in slot cars. Remember them? A small weight is attached to one side (not both) of the shaft, so that when it spins the weight, being off balance, off center, causes it to start shaking, vibrating. The assembly is wrapped in plastic foam material to make it silent.

Pager Number

With nothing more than the phone number, there are ways to find your capcode and frequency. The first involves a little social engineering—calling other numbers that are close to yours, from a pay phone that accepts incoming calls. Sooner or later, someone will respond, at which time the spy identifies himself as a technician for the Tree Frog Paging Alliance, calling to advise you that the system will be shut down for maintenance later that day.

"What? Tree Frog? Hell, my pager isn't on Tree Frog. I never even heard of them."

"Oh, well, you see, there have been some mergers and reassigning of blocks of numbers recently (you should have received a notice in the mail), and Tree Frog is taking over maintenance for several other paging companies. Which one are you using, incidentally? Our database is being switched to the new Oracle system and everything isn't transferred yet."

Now, you may smell a rat and get suspicious, but if the spy doesn't get the info he needs from you, he will keep trying and sooner or later, someone will give it to him.

Next, he can start calling all of the paging companies and explain that he found a pager but there are no markings on it except a Dymo label with the pager's number. He can't call it because the battery is dead. Ask if the prefix is one of theirs, implying that if it is, you will mail it back to them.

Another method, used only as a last resort, is for the operative to page you and leave a nonworking number with a few digits following it, then set the software to flag that sequence of digits. This is hit or miss and may have to be repeated many times, on many frequencies, before it succeeds. This should tip you off that this is what is happening.

Once the spy has your capcode programmed into his system, he has a complete list of every alphanumeric message you receive and every number left for you to call with numeric-only pagers. Over a period of time, he can find out who has some of these numbers, what kind of business they are in, and perhaps some details of their personal lives. Bits and pieces, gathered over a period of time, could eventually result in a rather detailed and sophisticated dossier. On you. So, as you can see, while it takes a certain amount of determination and expertise to single out a given pager, it can be done.

Now, as to someone intercepting messages at random, as they appear on the computer screen, with a little social engineering a spy could determine who it is assigned to. Who you are. From there, he can do the same thing—put together a great deal of information. This is possible. It could happen to you. It is also possible that you might be struck by lightning. Chances are about the same. Unless the spy finds something interesting.

Mama pages you to complain that Junior didn't take the garbage out and to bring home a bag of kitty litter. Ho hum.

Sadie pages Clem telling him to "git yore ass home right now."

Yawn . . .

Your sweetie pages you. "Oh, baby, I can't wait to ravish your body. Hurry home my lovey-dove stud." Lovey-dove stud? Your ratings just went up.

Think about it. There are people out there intercepting pager messages. Government agents. Industrial-strength spies. Kids with scanners. Think about it.

Big business hasn't done a helluva lot to provide privacy and secrecy of your pager messages. I called a number of vendors listed in the Pretty Good (Yellow) Pages. An hour wandering through the labyrinth that is corporate voice mail. Eventually, I got a few answers to my question: "Do you offer encrypted paging service?"

- Sprint: No.
- SkyTel: Unknown. After fifteen minutes of being bounced around through their voice mail system and being asked for a PIN that I do not have, I gave up. One more down and counting.
- Airtouch: Voice mail again. Receptionist did not know what I meant by "encrypted" and said they "do not have a support group." I gave up.
- PageNet: No.
- MobileComm: No.
- AT&T Wireless/MetroCall: No.

And Then Along Came (Uncle) Sam

The same thing that happened with cellular radio a few years ago is happening with the paging industry. Word has gotten out that pagers can be monitored. Several years after the horse is out of the proverbial barn. So, the government is doing something about it. They are passing laws that say people cannot listen to, monitor, certain frequencies. The equipment—the decoder, the software—is coming under heavy fire. There have been several arrests of persons who distribute the decoders. More arrests will follow. More equipment will be confiscated because the government doesn't want people to have it. This will put a damper on unlawful interception of pager messages. But for now, there are thousands of copies of the software and hundreds of thousands of scanners.

How to Defeat These Systems

Eventually, most of the paging vendors will come up with a secure system—secure against kids with scanners and probably against industrial-strength spies. But not selected federal agencies, of course. They will not allow people to use anything that they cannot defeat.

- *Probably the easiest way is to get a digital PCS phone*; they have a built-in pager. True, it is more expensive than a pager, but if you already have an (obsolete) analog cell phone and a pager, the overall cost may be less. Naturally, you will have to change your pager number and get new business cards, so you have to decide whether or not secure paging is important enough. PCS will completely lock out the casual listeners. At least for now. It appears that PCS isn't as secure as it is advertised to be; while it is digital it is apparently not encrypted. At least not the voice transmissions. So, sooner or later someone will build a decoder that converts PCS into plain speech. And the game goes on.
- *Another option is to use the V-One system.* According to their Web site, the messages are encrypted using RSA's RC4 encryption with 128-bit encryption keys. Therefore, the messages could not be decoded and would be illegible and meaningless to the hacker. Cloning would be virtually impossible since the pager's secret encryption key along with its capcode would now be required in order to cipher any intercepted messages. The pager's secret encryption key is not transmitted over the air and therefore could not be intercepted illegally. Moreover, access to the secure folder where secure messages are stored on the pager device is protected by a user-defined access password.

Finally, someone is doing something:

V-ONE Corporation
20250 Century Blvd., Ste. 300
Germantown, MD 20874
www.v-one.com/vpn_news/1997/news-08-28-97.htm

HAS THY NEIGHBOR BUGGED THEE?

Now, let's rehash some things—look into how you can find out if someone has you under surveillance. Remember about probability? About motive, method, and opportunity? Let's start with motive. When someone does something to another, he generally has a motive. A purpose. It may be very vague, without any logical reason, and it may just be compulsion. Someone might bug you just because you happen to be in the wrong place at the wrong time. This is one reason that you might be under surveillance. Improbably, yes. And also unlikely that such a person has any use for the intercepted information, unless you are blabbing about the Krugerrands you have buried in the backyard. Didn't think I knew that, did you?

Now, as far as you being bugged by design, specifically, by intention, let us look, again, at some of the reasons why this might happen. Who are you, and what are you involved in that would cause someone to want to overhear your conversations? An attorney working on a civil action or a corporate merger, in which millions of dollars are at stake? Perhaps an advertising executive who is finalizing the plans of a massive ad campaign for a new product in an industry where there is a great deal of competition? Someone who law enforcement believes is involved in illegal activities such as drug dealing? Even though you are not.

Do you have, for whatever reason, enemies who would like to "get the goods" on you? Remember about being in a relationship that you know is coming to an end, and there are things such as child custody or a property settlement? If so, then there is always the possibility.

As to opportunity, who can bug whom depends, in part, on who you are. You and I could never bug the Oval Office. No way could either of us even get on the grounds of the White House. But Nixon did. Remember the 18 minutes of Watergate tape? Opportunity.

Your husband or wife, obviously, has access to your bedroom. But if you leave your front door unlocked while you are both at work, so does anyone else. If you have your home well secured (a sophisticated alarm system, Medeco locks on the doors, a German shepherd in the backyard, and neighbors that look out for each other, etc.), then it is unlikely that anyone will be able to hide a bug inside your mattress. Should you have reason to believe that someone wants to spy on you, keep this in mind.

Now, understanding the above, look for the signs of surveillance.

Do you remember what you read back at the beginning of this book? Let's go over this again.

Are there people who seem to know things they are not supposed to know? Do they sometimes let something slip—say something and then try to cover it up or change the subject to draw your attention away from what they said, or started to say? Do some people always seem to know what you will do next and are prepared for it? In a competitive business, are there other companies that repeatedly underbid you by a very small margin? Has your Aunt Martha apparently discovered the plot to play *Arsenic and Old Lace* and asked her attorney to draw up a new will, cutting you out completely? So much for motive and opportunity. Now a look at method.

Where would you go if you were a bug?

So far you have learned a fair amount about surveillance transmitters, but from a positive aspect. Much has been said about using and installing them, antennas, etc. Now let's look at bugs from a defensive standpoint. Finding and dealing with them.

First of all, to find something it helps to know what you are looking for.

All of us misplace, or lose, things now and then. More often than not, they are where they were supposed to be in the first place, or else in plain sight. Poe's "The Purloined Letter" . . .

So, while you are searching, picture the lost item in your mind. The size, shape, and color.

So it is with searching for listening devices.

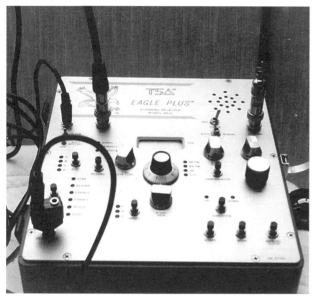

The Eagle Plus in operation on a sweep.

What does a bug look like? Above, I mentioned some of the things bugs can be disguised as, the products from PK Elektronik, for example. That's what a bug looks like. One transmitter, from AID, looks a bit like a 35mm film can, about the same size and shape. Others are bare printed circuit boards or little black boxes with the circuit boards inside. That's what a bug looks like. A bug can look like virtually anything, and it can be hidden inside damn near anything.

USING A BUG DETECTOR

A bug detector can be anything from a simple field strength meter to a countermeasures receiver to a $75,000 TekTronix spectrum analyzer. They all do the same thing—detect RF energy. But there the resemblance ends. The spectrum analyzer is the best overall device for finding surveillance transmitters but it is not cheap. An FSM can be made for a few bucks but is not recommended for serious searching.

Countermeasures receivers include the Scanlock, the Eagle, several models from Marty Kaiser, the TD-53 from Capri, and others from REI and ISA. Some are better than others, but—well, the great Ansel Adams once said something to the effect of, "Even the simplest of cameras is better than the greatest of photographers." Whatever you use, learn it well.

While the expensive systems such as the Scanlock and the Eagle will find almost any RF bug, so will some of the less expensive types. And while it is definitely a good idea to use the best equipment you can possibly afford, the skill of the operator is also important. Skill, the willingness to learn, and determination to do the job right.

The professional-quality receivers offer bells and whistles. Some of them

- have greater sensitivity
- can be programmed to block out signals from commercial radio and TV stations and others that could cause false readings
- have a wider frequency coverage
- have a direct frequency readout on the panel
- display the type of transmission (AM, FM) on a screen
- are more rugged and stand up to heavy use in the field
- can get the job done in less time

MYTHS: It is written that when a surveillance transmitter is operating, it is possible to locate the listening post by tuning in on the local oscillator (LO) of the receiver used to monitor the bug. In the old days of tube-type equipment, this was possible, and in fact was done with much success. But with solid state equipment, this is no longer true. The LO signal is just too weak to be detected at any distance. See for yourself. If you have a tube-type FM radio, tune it to the high end of the band: 108 MHz. Then take another, transistor radio and tune slowly through the band between 95 and 98 MHz and at the very bottom of the scale. You should hear a loud tone that is the LO.

Try the same experiment with two transistor radios and you probably won't hear the tone at all.

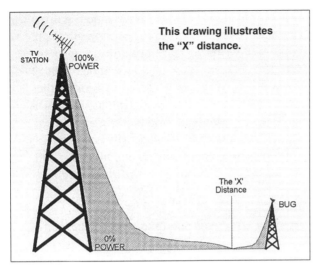

This drawing illustrates the "X" distance.

TV STATION

100% POWER

0% POWER

The 'X' Distance

BUG

Inexpensive transmitter detectors use meters or a series of LEDs to indicate the presence of a radio signal, but you have no way of knowing what that signal is. It might be a bug, and it might be KBUG-FM. And more often than not, a thorough physical search won't find anything because metal objects, desks, filing cabinets, venetian blinds, etc. rebroadcast strong signals from commercial stations. No one who is serious about finding a surveillance transmitter would use such a device. It's a waste of time.

The absolute least I would recommend is the Capri TD-53. It goes for about $500 and is a pretty good instrument for the price range. There is also the Model 262 from Sheffield, which is about the same price and is probably at least as good as the TD-53, but I have never used this particular device. If you can afford it, get an Oscor from Research Electronics, Inc. in Tennessee, and learn to use it well. This is an excellent product. If money is no object, get a spectrum analyzer and learn it well. Whatever you have, use a test transmitter. Have someone hide it and then practice finding it. Do it again. And again . . .

When you do a search, use headphones instead of the built-in speaker. When the device finds a bug, it will start oscillating, making feedback. A listening spy will hear this and you will have tipped him off that you found the bug. Bad move.

Depending on the bug's frequency and the amount of RF from outside sources (broadcasting stations, etc.), the detector may have to be within a few inches of the bug for the unit to find it. This is because of something called the near field distance. In the Scanlock manual, this is called the "X" distance. There is a point, in the distance from a hidden surveillance transmitter, where its signal will be received by the bug detector at a higher signal strength than another signal. No matter how powerful it is.

Some bug detectors receive their entire range all at once rather than being tuned across their coverage like a conventional receiver. The diode and "op-amp" front end types work this way. Other systems scan through their coverage over a period of time. This can be from a few seconds to several minutes. Some have different scanning rates, such as the Eagle, which has three.

In a professional sweep, just before the search area is entered the team goes silent. No talking whatsoever. The equipment is set up without making any loud noises. This is to avoid alerting anyone who might be listening. Once that first, silent, phase is complete, the physical search begins.

Mister Whidden

The Eagle, for example, which was developed by Glenn Whidden of Technical Services Agency, is first put into record mode, in which it will scan through its entire frequency coverage. It will intercept every

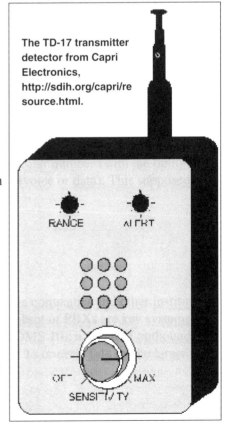

The TD-17 transmitter detector from Capri Electronics, http://sdih.org/capri/resource.html.

RANGE

ALERT

OFF   MAX

SENSITIVITY

transmission in the area, and the frequencies of these signals will be stored in one of its two memories.

This is done twice, the first time outside the area to be swept. Then it is repeated inside the target area, where it will ignore the signals it has stored in memory from the first time. This eliminates false signals such as radio and TV stations, police and taxicabs, cellular phones, etc.

On the second scan, it stores anything else it hears in its second memory. Once this is done, it can "dump" a list of these frequencies to a portable computer. Then they can be methodically eliminated as surveillance devices. What cannot be eliminated is further investigated the "day after." This usually involves several hours of carefully tuning ICOM receivers in the back of a cramped and sometimes hot van. The Eagle is set up on a table or desk and connected to various other equipment; it is not truly portable.

The 2050 CA Handheld RF Detector.

Other types, such as the Scanlock, the TD-53 from Capri, and the Model 262 from Sheffield, are carried by the operator who searches the area, poking the antenna (which is connected to the unit with a short cable) into every corner, every bookshelf and potted plant, item of furniture, etc. in the room. The antenna is held at different angles (right, polarization—you remembered!) and adjusted to different lengths to make sure it picks up even the weakest of signals.

So, again, to do the job right, use the best equipment you can get, read the manuals carefully, practice, make the search thorough, and don't overlook anything. You never know . . .

The Optoelectronics Xplorer

People come up and ask what is that thing I am holding, the most common question being, "Hey, is that a scanner?" "Well, not exactly . . ."

The Xplorer is a test receiver that sweeps through its entire coverage of 10 to 2,000 MHz (except cellular radio) in about two seconds, displays the frequency, and demodulates the audio. It also can display DTMF and PL (private line) tones, store 1,000 frequency hits, and a bunch of other things. A fascinating product—one of the best from Optoelectronics, which has a reputation for quality equipment. That's what the Xplorer is.

Its uses include being able to read the frequency of a two-way radio being tested in the shop, or in the field, or to pick up the FM radio transmissions of any radio that happens to be within range. Technically, it is known as a "near field" receiver, meaning that it intercepts the strongest signal it hears. The typical distances, from the operator's manual, are:

Cordless phone	10 mw	49 MHz	10-15 feet
VHF radio	100 mw	150 MHz	600 feet
UHF radio	100 mw	450 MHz	1,000 feet

The numbers depend on many things, starting with the "X" distance explained a few pages back. If you are near a powerful station such as a pager transmitter, it will not pick up anything else until you press either the "skip" or the "lockout" button. Antennas are also very important with the Xplorer, and several are available. There are also filters:

- high pass—lets signals above a certain frequency through and blocks those below
- low pass—just the opposite of high pass
- band pass—lets signals within a certain range through and blocks all others
- band stop—blocks a particular range of frequencies and lets the others through

The 2065 Under Desk Mount RF Detector.

The 2060 LT Wearable RF Detector.

The Xplorer is useful in finding surveillance transmitters due to its speed, wide coverage, and portability. Use it like you would any other portable detector. This is also true of FM body wires, since it is small enough to conceal easily in a coat or jacket pocket.

Mister Kaiser

Marty Kaiser started his company in 1965 and, in the years that followed, designed and produced dozens of products that have come to be known for their simplicity, ease of operation, and high quality. Marty is a true electronics genius who has never formally studied the subject—never taken a single class. He is entirely self-taught.

Starting off as one of those kids who would rather tinker with radios than play baseball with the other kids, he has built a company known the world over for excellence in its merchandise. Here, we will have a look at some of his creations with photographs and specifications. If you have Internet access, you can learn more by visiting his Web site at www.martykaiser.com. There, you will find more detailed descriptions, updates, and, for some products, the complete operator's manuals. I will be updating the site as products are improved or new ones developed.

The 2050 is an ultra-sensitive handheld RF detector that indicates the presence of a signal by an audio tone that varies in frequency in proportion to the strength of the signal. (This can be switched off when desired) and a reading on the panel meter. It also demodulates the signal so that the operator can hear it through the headphones, which are supplied with the unit.

The 2050 has five switchable bands, each of which can be tuned through manually by using the knob on the right side of the case:

- 1.5 to 6 MHz
- 6 to 20 MHz
- 20 to 50 MHz
- 50 to 100 MHz
- 70 to 400 Mhz
- The "B" band, which covers 10 to 1,000+ MHz. Sensitivity is such that a small transmitter of 1 to 10 mw will be detected at a distance of about 5 to 25 feet. But keep in mind what you have read elsewhere in this book, remembering that this will depend upon background (ambient) RF and other factors.

Now here is a clever gadget. Remember the "inventor"? If the lawyer had installed one of these under his desk, the spy may well have been discovered the instant he walked in the door. Then the lawyer could have had him followed and later dealt with in any of various ways.

This is a sensitive broadband field strength indicator that covers 1 to 2,000+ MHz and runs on a single 9-volt battery for some 500 hours. It can be mounted anywhere as long as the panel meter is easily visible to the owner, but no one else. Under a desk, of course, or in a desk drawer, or wherever else.

The 2060 Wearable RF Detector is a small, sensitive broadband detector covering 1 to 1,000 MHz that can be hidden in a shirt pocket with the flexible antenna concealed inside the sleeve or hung over the shoulder.

Then the sensitivity control is slowly adjusted to where the indicator light comes on and then backed off to where the light goes off. It is now set at its most sensitive position for the background RF in the immediate area. In the presence of an RF signal, it will vibrate and the LED will turn on and then off again when the signal disappears. This way you can tell a continuous signal, as from a bug, from that of a commercial two-way radio such as in a taxi or police car.

A little more "James Bondish," here is the 2055, which will detect not only the RF from transmitters, but also from computers and even the LO from radio receivers at a very close distance. From the manual:

> One of the two antennas detects toward positive (on the panel meter) and the other toward negative. Therefore, if the source of the RF is far enough away, about 20 feet, both antennas receive essentially the same amount of energy and the needle remains at zero. As the ratio of the distance between the RF source and the antenna decreases, one antenna receives more RF than the other and the meter swings in that direction. This effect is known as differential detection. Location of the RF source can be pinpointed by viewing the meter, but this is not required as the tone changes in the headphones do the same thing, freeing the operator to view the area being searched.
>
> The 2055 covers frequencies ranging from lower than 10,000 Hz (10 KHz) to over 1,000 MHz or 1 GHz. The system consists of the amplifier/indicator assembly, RF head assembly, head extension rods, head extension cable, two multi-section antennas, two right angle BNC connectors, special high-impedance 2,000-ohm headset, carrying case, and instruction manual.

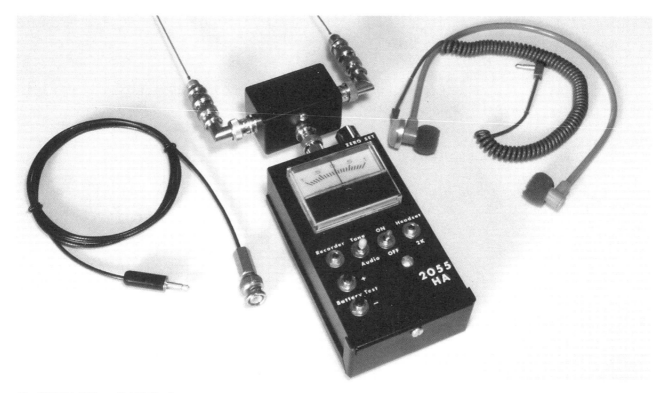

The 2055HA Differential RF Probe.

The 2044 RF Detector.

THE BUG BOOK

Fascinating. Would like to have one of these myself!

Power is from two 9-volt alkaline batteries.

The 2044 ambient radio frequency level detector indicates the presence of many types of RF, carrier current, and infrared light beam transmitters. The unit consists of the main amplifier/detector assembly, a multisection RF antenna with extension cable and probe, an optical probe, and a carrier-current probe for use with AC and telephone lines. Three methods of detection are accomplished with the 2044: audio feedback, relative signal strength, and audible/visual alarm. An excerpt from the product information sheet:

To use the 2044, plug the appropriate probe into the front panel jack, turn the power on, and carefully sweep the area with the probe. For RF detection, expand and collapse the antenna for different frequency ranges while listening through the built-in speaker. In RF mode, a background hiss will be heard from local commercial broadcasting stations, so adjust the gain (volume) control to a comfortable listening level. If a transmitter is present, it will cause feedback (a squealing audio tone) and by adjusting the volume as well as holding the probe at different angles, you will be able to zero in on it.

Detecting infrared devices is done the same way, keeping in mind that the transmission may be more directional so you need to be more diligent in holding the probe at many different angles.

To use for carrier current devices (CCD), the 2044 is plugged into the AC power line and the telephone lines and the volume control are adjusted, listening again for feedback.

Since this is an ambient type device, it can also be used as a sophisticated transmitter detector similar to the 2065, but it is much more sensitive and effective.

The 2057A RF Locator.

COUNTERMEASURES

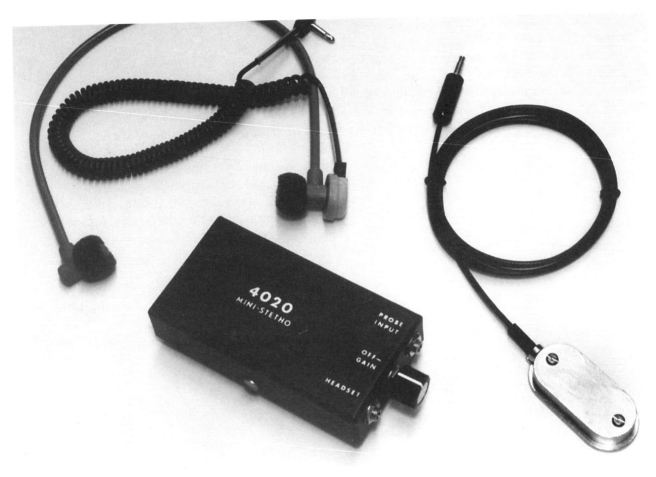

The 4020 "Mini-Stetho" Electronic Stethoscope.

The 2044 is powered by a rechargeable gel-cell battery pack that will operate it for about 50 hours of continuous use. It is shipped complete with all probes, headset, carrying case, battery charger, and operator's manual.

This is an ultra-sensitive 10 to 2,000 MHz radio frequency locator with audio and visual readout from the front panel meter. It demodulates (detects) both narrow band and wide band FM, AM, and AM/FM subcarrier signals. It can be used in either of two modes: in tune mode, where the operator manually adjusts the unit through its entire range, or in lock mode where the device automatically locks on to the strongest signal present, regardless of frequency. It has a narrow band-pass filter to further enhance feedback detection.

The 2057A is completely self-contained, operating from a built-in rechargeable nicad battery pack and has an LED indicator on the panel to indicate battery condition. The 2057A is designed to be used in a single location but can be carried easily throughout the area to be searched. Once done, place it in a convenient location and adjust it for the background (ambient) RF level.

The 2057A is shipped complete with antenna, battery and charger, carrying case, and operator's manual. And now, back to the "James Bond" stuff—here are two fascinating products.

The 4020 is designed to detect conducted sound from clocks or mechanical trigger devices such as may be used in explosive devices. Bombs. And this it does very well since it is purchased by a number of local and federal law enforcement agencies.

The probe input jack is designed to accept transducers (e.g., microphones) with impedances ranging from 1,000 to 500,000 ohms. The headset jack accepts headphones with impedances from 16 to 5,000 ohms but is specifically designed for the one that is supplied with the unit.

To operate, plug the headphones and probe in the appropriate jack (making sure not to get them reversed), place the probe against the surface of the device or location to be inspected, adjust the gain control, and listen for suspicious or interesting sounds.

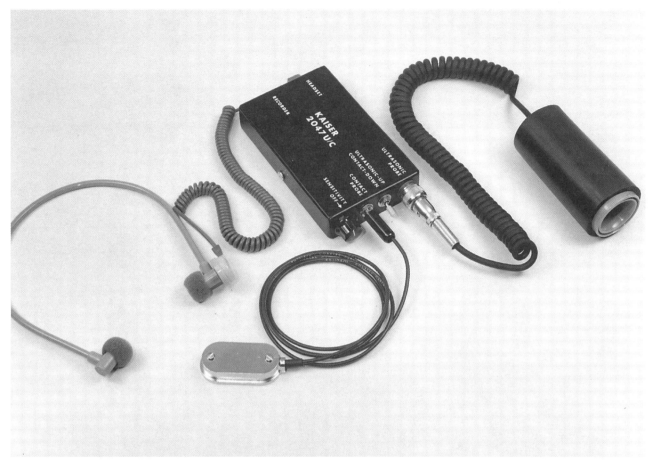

The 2047U/C Stethoscope Amplifier assembly.

It is supplied with the amplifier assembly, contact microphone, special high impedance (2,000 ohm) headset carrying case, and operator's manual and operates for about 100 hours on one standard 9-volt alkaline battery. Dimensions are 1 by 2 by 4 inches.

The 2047 detects mechanically transmitted (conducted) sounds in the audible range as well as airborne sounds above the normal hearing range (ultrasonic) from 38 to 44 KHz.

To use as a standard stethoscope, insert the contact probe into the proper jack, connect the headphones (supplied), place the probe against the surface to be tested, and adjust the gain control.

Operation in ultrasonic mode is the same except that the probe is seeking signals from the air rather than from a surface. To test it, have someone tear a sheet of paper in half at a distance of 20 to 30 feet, which generates ultrasonic sound. It should be heard clearly through the headphones.

It is supplied with the amplifier/detector assembly, contact probe, ultrasonic probe, headset, carrying case, and operator's manual. Power is from a single 9-volt alkaline battery, which lasts about 70 hours.

Optional accessories include:

- EC20C: 20-foot extension cable for contact probe. Up to four can be used.
- EC20U: 20-foot extension cable for ultrasonic probe. Use up to three.
- SM80-6: 6-inch spike assembly for use with contact probe. For testing soil samples, i.e.,, potted plants.
- SM80-12: 12-inch spike assembly for use with contact probe.
- DM80C: Dynamic contact probe
- 2101 Doppler Probe: Ten GHz radar probe that detects mechanical clocks or timers through soft-sided packages such as attachés, suitcases, wooden or cardboard boxes, etc.

The 1019 Amplifier/Preamplifier.

The 1019 was designed as a multiple input/output low-noise preamplifier/amplifier that can provide a near-perfect match to just about any microphone or recorder. It includes the preamplifier, a half-watt (500 mw) power amplifier with built-in speaker, tone generator for testing amplifier performance and unknown lines, switchable excitation voltage for external accessories, and a low-pass/high-pass filter selector.

The way this applies to surveillance is that when recording an intercepted conversation, using it will improve the quality by ensuring that the output from the receiver matches the recorder.

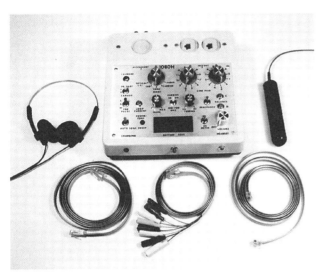

The 1080H Telephone Analyzer.

Dimensions are 2 1/2 by 3 by 5 inches. For those who don't want the power amplifier, get the 1059 preamplifier. It's about one quarter the size.

The 1080 is a sophisticated telephone line analysis system that can be used to perform a multitude of tests on a phone line. The front panel may seem confusing at first glance, but everything is explained clearly in the operator's manual. Following are explanations of some of these controls. The two rotary switches on the right are used to select all possible combinations of the two input jacks at the top right and will detect a tap using the phantom pair method. They also test the line balance between tip and ring and ground. They are used along with the two push-button switches C and D.

The rotary mode switch on the left selects the main functions of the analyzer. AUDIO monitors any audio present on the line. (You will read more about this later in the book.) CARRIER selects the mode for checking the line for a CCD. TONE generates four swept audio tones that cover all possible combinations of DTMF and other audio frequencies between 200 and 4,000 Hz. (Incidentally, the bandwidth of telephone lines is 300 to 3,000 Hz. So this instrument will also detect "out of band signaling" tones.) TDR is the Time Domain Reflectometry mode.

The loop-current push-button switch places a 100-ohm resistor across the pair selected with the line pair rotary switches. The meter reads the voltage drop across the resistor and thereby calculates the line current. The resistance push-button switch measures the DC resistance of the line selected. A conversion chart is included in the operator's manual. There are three binding posts on the back for connection to ground and to an oscilloscope which is used with the TDR mode.

It takes a little practice, but the operator will soon discover that the 1080 isn't so complicated after all. The 1080 is shipped with all necessary cables and connectors, an infrared probe, headphones, a gel-cell battery and charger, carrying case, and operator's manual.

What the TDR does is send a signal down a pair of wires or a cable in such a way that if there is an opening, a break, a short, or a splice (such as a wiretap), this will cause the signal, or part of the signal, to be reflected back to the device. The interval (time) elapsed is electronically converted to distance so that the tap (or break, or whatever) can be located easily.

Standard TDRs are very expensive, with some TekTronix models well into five figures. The 1080 will accomplish the same things for a fraction of the cost, partly because it uses an external scope and does not have a printer.

As stated in the manual:

> The 1080TDR, when used with an external oscilloscope, analyzes wire pairs or transmission lines. The TDR is essentially an echo ranging device. It generates a short, very fast rise time pulse that travels along the wire at a speed determined by the velocity factor of the wire. When a discontinuity is encountered, the pulse is reflected back along the wire pair to the 1080TDR and oscilloscope. The actual distance to this discontinuity can be measured by a simple calculation.

The mechanics of using a TDR are simple. Interpreting the results are not. OK, true—one can do the exercise in the manual where a 100-foot length of phone wire is tested and the measurement calculated, but learning to recognize the subtle differences in waveforms that result from different conditions takes a lot of experience.

The 1080 is shipped with cables for connecting it to a phone line (RJ-11) and scope (BNC) carrying case and operator's manual.

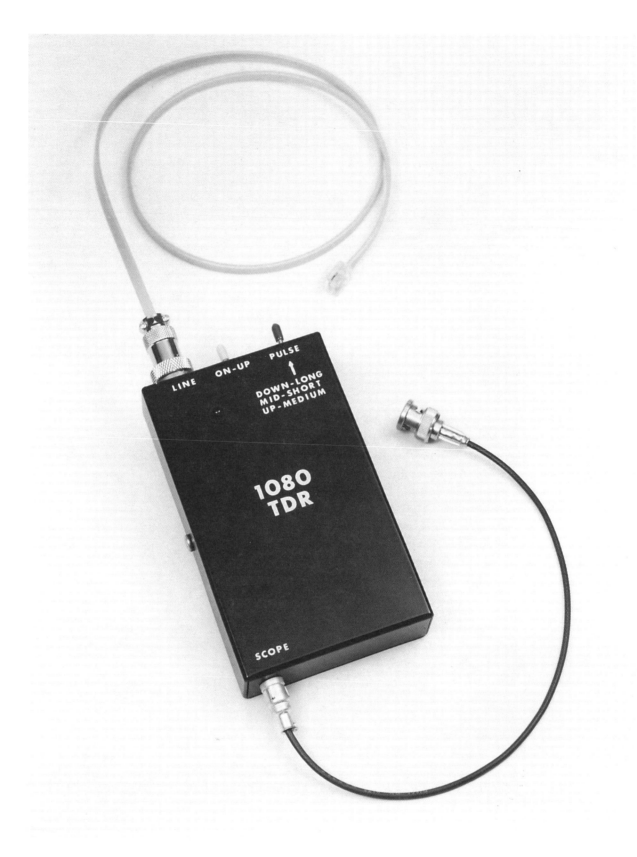

The 1080 TDR Time Domain Reflectometer.

The 2059H Tape Recorder Detector.

The 2059 very low frequency (VLF) detector covers from 5 to 200 KHz. Many, BUT NOT ALL, tape recorders contain a bias oscillator that operates in this frequency range. (Bias is used to align the small magnetic particles on the tape, called "domains," previous to recording.) This radiation can be detected at distances of from about 6 to 60 inches. The 2059 indicates the presence of a bias oscillator radiation via a small signal light (LED) or vibrator.

The unit can be used in "sweep" method similar to an RF detector, by passing the loop antenna through the area or over a device suspected of concealing a recorder. This could include a person, but in such a case it can be used covertly by placing it in a pocket or inside a hollowed-out paperback book, which might get it a little closer to the person suspected of having the recorder.

If you use the 2059 you should be aware that, as stated in the manual, not all recorders emit this bias signal—such as the "tapeless" types that store digital audio in memory chips and models that use a permanent magnet. Also, be aware that the Nagra, which the feds use, has a switch that turns the bias off when being used for covert recordings. Also, TV and computer monitors may interfere and give false readings. The monitors should be turned off if possible, and the TV donated to a homeless shelter.

The 2059 is powered by two standard 9-volt batteries that provide about 40 hours of operating time.

The SCD 5 is a battery-powered device that detects VLF (10 KHz to 700 KHz) signals that are used by wireless intercoms and wireless extension telephones that send their signals through the power lines. As you will recall, these devices are sometimes misrepresented as being secure, but they are not. They can be intercepted by anyone who uses the same pole transformer, which may be several houses or an entire apartment building.

To use it, just plug the appropriate cable in, set the volume and gain controls, then select one of the two range settings and tune through the coverage with the panel knob. When a signal is detected, the unit will lock onto it (at several places on the dial). Then adjust the volume and listen.

The SCD 5 is powered by two 9-volt alkaline batteries and is supplied with cables for the phone line (RJ-11) and the 110-volt power line.

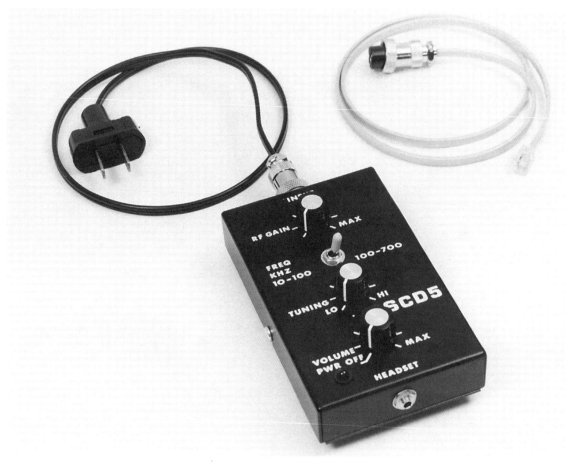

The SCD 5 Carrier Current Detector.

The 1040A Basic Countermeasures Equipment Package has all of the tools for radio frequency and audio surveys in a standard attaché case with foam insert. Included are the following:

- 1059 general purpose amplifiers with headset (two each)
- 2050CA RF locator with visual and tonal readouts
- 2030 carrier current probe
- 2040B RF and optical test oscillator
- microphone, general purpose test microphone
- SLR9A/AUT9 ultrasonic/audible line tracer
- 1040-4 hot-pack
- VOM multimeter
- IR probe infrared optical probe
- lantern, adjustable focus (two each)
- radio: special AM/FM radio
- cables: miscellaneous patch cables
- adapters: cable adapters
- manuals for all included products

The 1040B is an intermediate level Countermeasures Equipment Package. It has all of the tools needed for radio frequency and audio survey (without the carrying case), including the following:

- 1059 general purpose amplifiers with headset (two each)
- 1019 power amplifier with loudspeaker
- SCD 5 carrier current detector
- 1059VLF direct conversion receiver (two each)
- 2045B feedback detector
- 2050CA RF locator with visual and tonal readouts
- 2030 carrier current probe (two each)
- 2040B RF and optical test oscillator (two each)
- SLR9A/AUT9 ultrasonic/audible line tracer
- VCD video countermeasures device (four each)
- 2047U/C ultrasonic/contact stethoscope
- 2062 RF and tape recorder detector (wearable)
- 1080H telephone analyzer
- 1080E electronic and modular telephone accessory
- 1080CT component test unit
- IR ptobe infrared optical probe
- cables: miscellaneous patch cables
- adapters: cable adapters
- manuals: for all included products

The 1040S has the same contents as the 1040A, except that a 2045B feedback detector is substituted for the 2050CA.

COMPUTER-AIDED SCANNING

CAS is the process of using a computer to control a scanner or communications receiver, enabling it to increase the memory channels to an unlimited number and scan through them at a much faster rate. To set up as many banks of frequencies as are desired, and be able to switch in and out from one to another at the touch of a few keys. This is fairly new, and to those of us who have an interest in scanners, a wonderful development in technology that makes it possible to do things that were not possible a few years ago.

Now, CAS, as it applies to countersurveillance, already exists, sort of, in several countermeasures receivers. One of the pioneers in this technology is Glenn Whidden who designed and built the ScanLock—for many years an industry standard used by professional sweep teams and federal agents. Later, he built the Eagle Plus, which also became an industry standard. We used the Eagle at a TSCM company where I worked as a sweep technician, and I got to thinking—why not use a scanner and CAS to accomplish the same thing? Well, almost. CAS isn't as efficient as the Eagle, and it is very slow in comparison, but it does have its advantages. The system can be programmed as you want it, and it is much less expensive. A good system can be put together for under a thousand dollars. Possibly for half of that if you can find some good buys on what you need. By comparison, the Eagle goes for a little over ten grand. To set up a CAS operation, you need the following four things:

- a computer
- a radio
- a program
- an interface

The Computer

Little need be said here as virtually any DOS/Windows computer will work—desktop, notebook, makes no difference. An old 286 will run s-l-o-w-l-y, a 386 will run fairly well, a 486 will run quickly, and a Pentium II really moves! There are also some programs that will run on the Mac, and, fortunately, more are being written.

The Radio

Scanner or communications receiver, it makes no difference as long as it can be computer controlled or can be modified for this. Many of the radios on the market today, including portable "handheld" models, have a

The 1040A Basic Countermeasures Equipment Package.

built-in RS-232 serial port that connects to the computer and are usually shipped with a software package that will make the system work. One thing to consider, though, is whether or not a particular radio has the frequency coverage you require. Does this include cellular radio? Listening to cellular is unlawful. It is also boring. But as there is the possibility of a cellular bug, it is necessary for a thorough sweep. The Eagle is not cellular blocked, nor is the Oscor. So you will need one of the radios made before the law was passed in 1993 prohibiting the sale or import of radios that tune cellular—or one made after that but which can be modified to "restore" cellular.

There are still a lot of the older models available, advertised on Internet newsgroups such as rec.radio.scanner, but some of these older radios are not equipped with a computer port. Meaning you will need to modify them, if they can be modified.

The Interface

This is what enables the radio and the computer to work together. It may be only a cable that connects them, it may require that you tap onto a connection inside the radio known as the discriminator, or "baseband" output, and it could mean building the Optoelectronics circuit board into the PRO-2006 scanner.

The discriminator is the point at which the incoming signal has been tuned in, converted to IF frequencies, amplified, and detected. From there, the signal feeds into the audio stage, amplifying the sound so that it is loud enough to be heard through a speaker. However, the signal, which is "pure" at this point, will become distorted as it is amplified. And this distortion will interfere with the working of the system. In many cases, it will not work at all.

Optoelectronics Revisited

Opto manufactures the OptoScan OS-456, for the PRO-2005 and 2006, and the OS-535 for the PRO-2035 and 2042 scanners. The kit consists of a printed circuit board that installs inside the scanner (note that this may void the warranty), the RS-232 computer cable with connectors, an instruction disk, and a manual. The manual (for the 456 I installed) was well done, with only one minor error that I caught, and is not difficult for someone with experience in soldering and general assembly work. A beginner might well be able to do this conversion with a little practice using a low-heat soldering iron and basic parts identification. The 456 was, at the time it was released, probably the best CAS setup that existed, and even now, several years later, it is still one of the best. As I have stated elsewhere, Optoelectronics makes only very high quality products. The software shipped with these kits includes a copy of Radio Manager, which is a good program and has continued to get better with every new version.

Some other nice features of OptoScan are that the 2006 can be connected to the Scout for reaction tuning: the 2006 automatically tunes to whatever frequency the Scout intercepts. And, if you have the Spectrum CD-ROM database from Per Con Corporation, the OptoScan will automatically look up and display the FCC record of the transmitter. A very nice product.

For more information on the PRO-2006 see http://home.ptd.net/~pro2006/ or www.optoelectronics.com, the Web site of the manufacturer of OptoScan and the Scout.

Has anyone ever officially defined "monitor"? Is it against the law to have a frequency counter that "intercepts" a cell phone frequency, even though you can't hear the conversation? Are there a certain number of words that have to be overheard to constitute "interception"? What if you are logging channels for research purposes, but have the sound turned off? Is that interception? Or searching for a hidden transmitter? Professional sweep teams use countermeasures receivers that tune most of the RF spectrum, including cellular. Is that a violation of federal law?

Suppose you have an old analog TV set that tunes the channels now assigned to cellular? Is that illegal?

The DEC notebook computer set up with a PRO-2006 scanner to run Radio Max. The old portable computer in the back is a Toshiba T-1000, which records numbers dialed from the line through the MoTron DTMF decoder.

Deltacomm

This is a complete system—with everything you need to interface the radio to the computer, including cables and software. It is available for the ICOM R-7000, R-8500 and R-9000, R-7100, and R-71A. The R-7000, as stated above, requires modifying for the S meter and installation of the CI-V computer (serial port) interface. Instructions are provided, but this requires experience. You need to replace a crystal filter and tap the baseband point.

Delta Research • P.O. Box 13677 • Wauwatosa, WI 53213

CAS Software

The programs shipped with radios are often "demo" versions. The idea is that you get to try them first and, if you like, you can buy or register them. Some are quite good, some leave a lot to be desired, and there are many to choose from.

Probe

Probe 2.0 is a DOS program (it will not work under Windows) for the Optoelectronics OptoScan systems installed in the PRO-2005-6, PRO-2035, and PRO-2042 scanners. I have not used this program, but I have read several reviews, all of which were very positive. Here are a few quotes I've seen on the Internet:

- The strength of Probe is in its ease of use.
- Probe is designed for set and forget, whether scanning or searching.

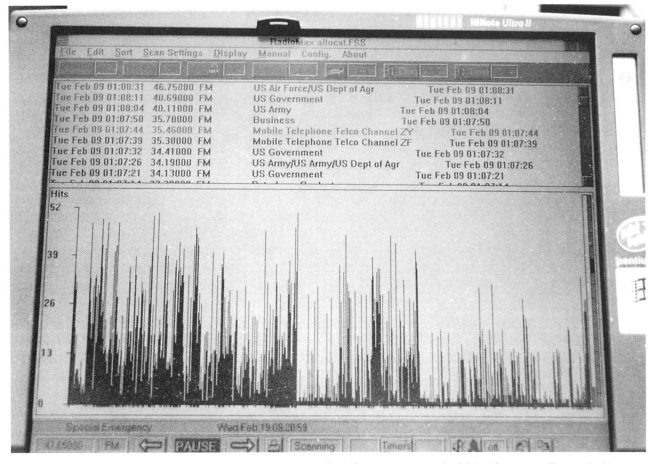

A closeup of Radio Max running on a DEC notebook computer. Each line represents an individual frequency. The vertical bar on the right is the signal strength indicator. The screen displays a group of frequencies that are included with the software.

- The informative screens provide a lot of information, most of which can be directly imported from the Percon CD
- Probe is designed to operate on almost any IBM compatible PC with minimal fuss. This package lends itself well to laptops and older computers.
- Overall, I am very impressed with Probe, and I use it almost daily. For computer control of a scanner, it is hard to beat.

Scanstar

Probably the most powerful and versatile CAS program available. It has many features including a spectrum display, unlimited frequency, and bank capacity, and it is capable of storing information settings for up to 10 receivers. It works on most popular radios and has output for squelch tone and DTMF decoders. It does have a fairly steep learning curve, so plan on spending a few days to master it, but once done I think you will find it an excellent application for overall CAS applications. Among the radios it works with are the ICOM R-7000, 7100, and 9000; OptoScan 456 PRO-2005 and 2006 conversion; various Yeasu models; and a lot more. To try it out, a demo copy is available free at www.scanstar.com.

Radio Max

Produced by Future Scanning, this is, in my opinion, the best program for countermeasures use, as well as probably the overall best for general CAS. It doesn't have all of the features Scanstar does, but it has many that Scanstar does not, costs about half as much, and is much easier to learn. Once it's installed and configured, you will be able to make it do what you want in an hour or so. The following is from the Future Scanning Web site:

Radio Max allows automated PC control of ICOM, OptoElectronics, AOR, Radio Shack, Uniden, and other receivers and scanners. Use the sound card to record and play audio directly even while scanning at full speed. Radio Max allows any PC running Windows to drastically improve the performance of radios and scanners, providing fully automated control of the radio allowing unattended scanning, logging, speech, and tape recorder control. Discover the ease and power of using a high-speed, graphics oriented receiver control program. Click and point or use the keyboard for instant control of almost everything. High resolution graphics constantly display ALL activity, position, locked channels, hits, real time squelch status, hit history, etc.

Radio Max supports automated time / date stamps, hits counter, autologging, history files, manual and automatic lockouts for birdies, priority channel, adjustable squelch trigger level, etc. Powerful scan delays and monitor time controls allow intelligent scanning preventing hangs and lost signals. Quickly jump, change direction, pause, slide up and down the tuning range with simple mouse movements. A tape recorder with remote control capability may be controlled by RadioMax, programmable by channel. Or, use the sound card to automatically record the radio's audio, using features built right into Radio Max. No need to fool around with "sound editors" or expensive add-on "options." Unlike some other scanning programs, Radio Max lets you hear recorded audio, and play it back, while scanning at full speed.

Installation takes all of three minutes, and without a long series of confusing questions to answer. Just select the directory where you want it installed, hit <ENTER>, and it's done. Next step is to configure it, by selecting the COMM port and type of radio you are using, and a few other little things that are clearly explained. That's it. Select one of the built-in frequency lists and it starts scanning. Fast.

Radio Max has features of special interest in countermeasures. It has Multi File (Bank) Scanning capability. This means you can enter the frequencies most likely to be used for bugs, depending on your personal situation, into as many different banks as you like, perhaps starting with those listed in the appendices. You could put old cordless in one file, 900 MHz cordless in another, etc. and then scan any combination of these banks (files), switching them in and out in seconds. Or, you can enter them all into a single bank. Or any combination thereof. To search a particular range, just enter the beginning and ending frequencies and a few other details and start it running.

Being able to instantly reverse direction is most useful; if you hear something and want to go back to it, just point the mouse and click. You can also pause the action with a single click. Everything is right there on the screen. There is no need to click on help files or struggle through pull-down menus; just point and click.

Two more very useful features are the time/date stamp that makes a permanent record, and being able to record the intercepted signals automatically on the hard disk. You can also use a tape recorder if you like. And if you later decide you want to make a copy, you can dump the file from the hard disk to the recorder through the sound card.

Saving the best for last, Radio Max displays all of the frequencies that you are scanning (in either a list or a range) on the screen. Each one has a vertical line that changes colors when the radio picks up a transmission. If it is something you want to listen to, just hit the space bar or click the <PAUSE> button, or, if not, click once at the bottom of the line to lock it out. That way the radio will not stop on that frequency again until you tell it to. A small bar underneath each frequency turns green to show at a glance that it has been locked out.

For unattended operation, you can set it to stop on each hit for any length of time and then move on; meanwhile, the program builds a "history" file, a permanent record of all hits. And not only that, you don't have to search through this hits file manually to see what is there. In seconds it can be converted to a scan file. In other words, all of the frequencies that have hits can be instantly made into a separate file that you can scan through. Sit back and let the radio do the work. Radio Max is an outstanding piece of software that anyone with a serious interest in electronic countermeasures should have.

OK, whatever equipment you have decided upon, once you get everything set up and running, you are still a few steps away from bug hunting. First— practice. Second—practice. Get to know the radio and the program so that you can operate the system without having to stop and think or look up something in the manual. Learn the frequency spectrum. There is a wall-size chart, approximately 24 by 36 inches, available from the U.S. Government Printing Office for about ten bucks. This is a good investment, as you can quickly look up a particular band and see what type of service is assigned there.

Now, make a plan. Think about what you have read in this book, as well as *The Phone Book*, if you have it. Think about the Introduction to this book— Who's (Bugging) Who(m)—and the type of transmitters that are most likely to be used under different circumstances. This helps narrow down the area to be searched first. Note: this does not mean eliminate any band or range of frequencies; it is a place to start. If you are an experienced scannist (scannerist?), you know well that such a search would normally take umpteen hours, days even, because the radio will stop on hundreds of signals that transmit continuously, and you would have to keep hitting the scan button endlessly to keep it going.

But with CAS, this is all done automatically.

What does a bug sound like?

OK, you say, enough about using these devices. How the hell will I know if I do find a surveillance transmitter?

Think about what you read on demodulating. If you are using a device that has only an LED or panel meter to indicate the presence of a transmitter, you will no doubt get false readings. So, you won't know for sure if you are detecting a bug or a strong signal rebroadcasted from a large metal object in the search area.

Now, if you have a demodulating type device, it is a good idea to start the search using headphones. Remember, if you use the speaker and come in range of a transmitter, you will get feedback, which will alert the spy, should he be listening at the time. So listen to the sounds you are making as you move around.

That's what a bug sounds like.

Or have a radio playing softly in the area.

That's what a bug sounds like.

If you don't detect anything then go to Plan B.

In a professional search, once the silent phase is completed, a source of very loud noise is used. A siren-like device that makes a whooping noise up and down the audio spectrum is fired up. This will make sure that the bug hears it, and so you will hear it. Useful in the case of a sound-activated microphone. Whatever the bug "hears" you will hear.

That's what a bug sounds like.

What about other types? AM, spread spectrum, single sideband, burst . . .

Well, AM will be much the same as FM, but there may be a faint "hiss" in the background, or static and fading as described in Chapter 1.

Sideband, again, is a little like Donald Duck. A communications receiver or scanner that has a BFO (beat frequency oscillator) mode is required to understand it clearly although an AM radio tuned very carefully may make some of the transmission somewhat understandable.

A burst transmitter you probably won't hear at all since the transmissions last only for a fraction of a second. Use a spectrum analyzer.

Now, digital spread spectrum may not be heard at all, or it may have different sounds. The Pac Bell DSS phone, for example. (Yeah, I bought it. But for research purposes only, you understand.)

The manual says that it is a "20 channel, digital spread spectrum 900 MHz handset with speakerphone," but there is no other technical information. So I sent several e-mails to the manufacturer asking for details:

- What are the base frequencies for the 20 channels?
- The manual index, under Channel Selection, refers to page 1 but that page has no such information.
- What, specifically, is to prevent someone else who has the same phone and is in range, from being able to intercept this phone?
- I don't see anything about pseudorandom noise generating or spreading code and anything else that indicated the user has a phone that is truly unique; no user settings. Please elaborate.
- What is the RF power output?
- Is it frequency hopping or direct sequence?

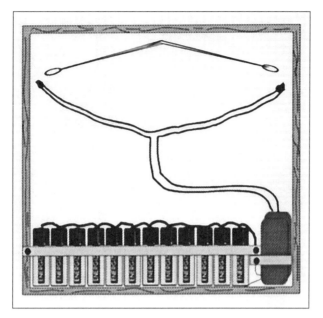

Using strips of nylon fabric and thumbtacks, you can hide a transmitter in the back of a picture frame in a few seconds.

None of these communications were answered.

The "900" MHz band is actually 902 to 928 and is assigned for amateur radio and certain low-power devices that are not required to have an FCC license. So, I plugged it in to see what would happen.

First, I used a field strength meter. With the antennas only an inch apart, the phone did register, but weakly, raising the needle to one on a scale of five. As a comparison, a 2-watt FM transmitter on the Family Radio Service band pegged the needle at a distance of two feet. So, an FSM might detect the presence of a DSS bug.

Then I set the PRO-2006 scanner to search that range in NBFM, made a call, and got signals on 907.03, 909.45, 915.84, 916.80, 917.76, 918.625, and 921.125. I repeated the search in wideband FM but heard no signals. In AM I got the same frequencies but the signal was louder.

The sound was a kind of whine of about 300 cycles that varied in intensity whether or not the phone was moved around. In the background, on some frequencies, were faint static and occasional clicking sounds. When I turned the handset off, the sound disappeared.

So, it is possible that you may detect the presence of a DSS bug with an FM receiver, at least the type used in cordless phones, but not necessarily all DSS bugs. Don't depend on this. Do depend on what you will learn in the next section.

THE PHYSICAL SEARCH

In the physical search, try to remember what you have read about installing transmitters in previous chapters. Learn to think like a spy! Also keep in mind that while the bug detector will find most RF transmitters, it will not find hidden microphones or other types of surveillance devices. The physical search will. Done right it will find virtually any inside eavesdropping device.

Appendix E includes a long list of things to check that you can use as a basic guide. It starts with a rehash of this chapter and other parts of the text, which I think you will find useful. As you look through it, take a look around the room you are in. How many things can you see that are not on the list? Probably at least a few at first glance. Now look again, more carefully, if you will, to see what else you find. Please keep this little exercise in the back of your mind when making a physical search.

The Grid
- Make a drawing of the area, called a hazard chart, showing possible hiding places.
- Draw a grid over the chart that divides the room into smaller areas, and check off each one as it is searched.
- Use self-adhesive inventory labels to mark the things that have already been checked. To be really thorough, have a second person make their own search, using a different color of label.

A small assortment of hand tools will be useful for taking things apart, removing switch plates and the baseplates of phones and lamps, etc., and a flashlight and small dentist's type mirror will be needed to peek into some places like underneath furniture. A good magnifying glass and a battery-powered ultraviolet light are also useful. The underside of drawers in furniture should be checked, and anything made of wood should be examined carefully for signs of tampering. Books should be removed from shelves and opened (they could be hollowed

The Hazard Chart

When making a physical search it is a good idea to make up a chart showing all of the possible hiding places for bugs. Then decide on a methodical search routine. You might start on the floor in one corner and work your way up to the ceiling, then along the wall to the next corner and continue that way. How you go about it isn't important. What is essential that you search the entire area very carefully. Do it right and you will find any listening device.

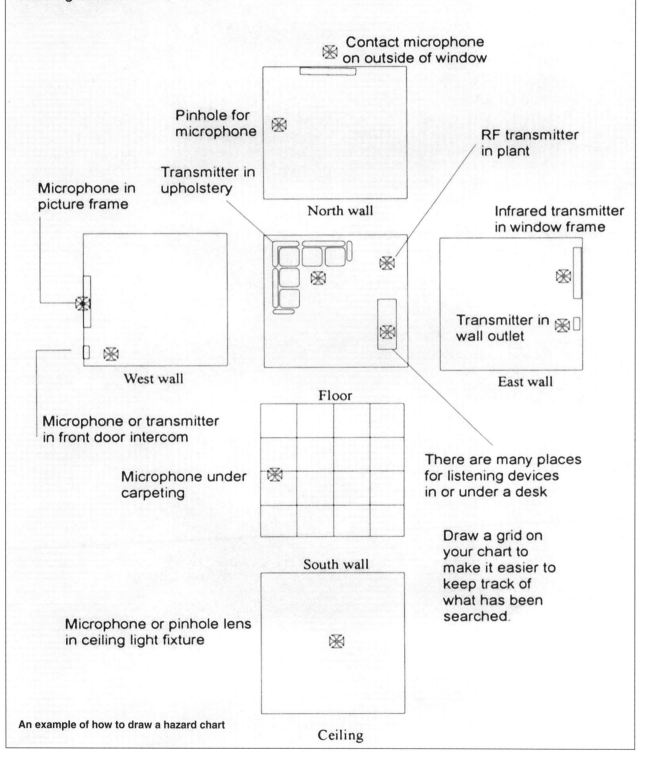

An example of how to draw a hazard chart

out), doors should be examined carefully from all angles, the top, and the bottom, since some are hollow.

The acoustic tile panels of drop ceilings should be looked at carefully for indications that any have been removed and replaced. Better yet, get a ladder, lift up one of the panels, and look inside.

Look for any indication that the molding around door frames has been removed, such as loose or missing nails or staples. Look for places on walls where the paint color doesn't match (use the ultraviolet light to see this more clearly), or pinholes that cast a shadow when a flashlight shines parallel to the wall.

Visualize an image of what you are searching for—anything that doesn't look like it belongs there.

If you are looking at the inside of a TV set or under the dash of a car, you expect to see wires. You should not see wires on a curtain rod or underneath a desk.

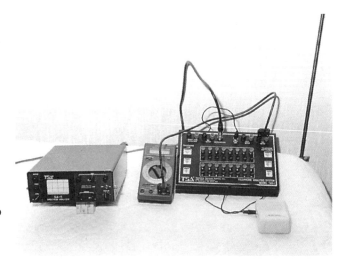

A telephone analyzer and spectrum monitor from ISA set up on a sweep. In the center is a digital multimeter, and on the right is a monitor speaker.

Look for printed circuit boards or small metal or plastic boxes, something connected to a handi-box, used where two pieces of electrical tubing (conduit or EMT) come together. Also think about batteries. Remember—any place that a person (spy) can get into, you also have to get into. It's hard, dirty work sometimes, but it has to be done right.

Finding Phone Transmitters

Finding a phone tap, whether it be a direct wire or an RF transmitter, can be more difficult than installing the tap. This is because a wiretapper needs only to find a suitable location along the line, but to find it you have to search every inch of the line from the phone to where it leaves the building. You don't know where it might be installed. But if you can access all of the wire, you will find anything connected to it. The search is done in two ways, the same as room audio transmitters—using electronic equipment and the physical search. But here they can be done at the same time.

While you use whatever type of RF transmitter detector is available, first you have to put a signal on the phone line so it has something to listen for. This you can do by calling a prearranged number where someone will leave the connection open for as long as you need. Place the speaker of a radio or stereo close to the phone so the sound will go out over the line, and listen for that sound in the headphones. You can also use a line trace device available from Marty Kaiser Electronics. And, in *The Phone Book* is a very detailed chapter on using an inexpensive ohm meter to measure the line resistance, which will reveal most phone taps.

Starting at the telephone(s), remove the cover and look for anything that shouldn't be there: wires with insulation of different colors than the rest of those in the phone. Connections that are spliced or soldered. Small metal or plastic shavings, or a brittle flaky brownish substance that is the flux residue from soldering wires. Tape of any kind. None of these belong there. Look for any kind of a circuit board with electronic components on it that doesn't look like the same material the rest of the inside of the phone is made of. If there is any doubt, just replace the phone.

Then start tracing the wire from where it is connected to the RJ-11 jack or connection block. Remove the box, or the plastic plate that covers it, and look inside with a dentist's mirror. From there, the wire might lead along the baseboards and around door frames or windows, or disappear into the wall or floor. All you can do is the best you can. If the wire goes down an elevator shaft, it will be hard to trace. If you are in a small apartment building, you might be able to follow it to where it leaves the building. Every situation is different.

You can also start the search from the other end, the 66 block or SPSP. Again, follow the wires as far as you can, looking for any other wires connected to them. Splices, small connection blocks, should also be also traced. Do you have extension phones in other parts of the premises? Start a new trace from there.

What If I Find a Bug?

Before we get into what to do, may I suggest something not to do: Don't holler at your partner, boss, client, secretary, etc. something like, "Hey, c'mere and look at this. It looks like a bug!" Don't say or do anything. Not yet. Just be cool.

What should you do? It depends, for one thing, on how you found it. If it was during the electronic search, then it is presumably still transmitting. If you found it in the physical search, which means that the electronic search didn't detect it, then it probably isn't still transmitting. But don't conclude this, because it isn't necessarily true. For now, assume that it is still live.

Use the bug detector to check as close as possible, use a frequency counter to determine the wavelength, but don't touch it. Even a light touch will be unmistakable to a listening spy, a dead giveaway. Leave everything exactly as it is until you decide how to handle the situation.

If the bug detector doesn't register it, it could be because it is transmitting on a frequency outside the detector's range. Normally, this isn't true; at a distance of only a few inches, the detector should still pick up harmonics or subharmonics, but it is a possibility. It could also be a remote-controlled device that has been turned off until you go away. The important thing is to try not to let the listener know you found the device. This may be useless, as said spy has probably heard the sounds made when it was discovered, but one does what one can. Another thing to remember is that these little critters tend to be sociable. Where there is one of them, there might well be more. And although one is dead, the others may not be.

TV surveillance revisited. I remember a TV movie, mentioned in *Don't Bug Me*. This was where someone placed a bug in the home of an FBI agent. (Ain't that a switch!) The agent found it because a neighbor told him that she had heard their conversations on her FM radio. The agent explained to his family that this was the mark of a professional: he'd placed one bug that was fairly easy to find to throw them off. The idea was that if they found it, they would assume that it was the only one. Bullshit. No competent operative would do this, and any professional TSCM team will assume that there could be another bug, and continue on. A good idea. Keep on looking. The search ain't over till it's over.

Once the search is complete, do several things.

Take pictures. Have someone else act as a witness, look at the suspicious wires, and take photographs. You will naturally want a close-up showing the block in as much detail as possible. However, if the pictures show only the block, it could be argued that they are of another block at a different location. So, you might start back far enough to include the background in the picture, then zoom in.

Take notes. Write up a record of the incident in as much detail as possible. Get everything down and be concerned about spelling errors and punctuation after you are done. Better yet, dictate it. Now, you have several options for dealing with what you have found:

- Disable the transmitter.
- Leave it in place and feed false information to whoever is listening.
- Say something that might draw out the spy and maybe catch him or her.
- Call in the FBI, if you are an established countermeasures technician, attorney, politician, or other "important" person. But the "average Joe" will just be written off as a nut case and ignored.
- Call the telco and tell them your phone is "tapped." Rotsa Ruck. You got your work cut out for you, as you read in Chapter 11 in the "Am I Tapped?" section.
- Have the device analyzed.

Secret Lives Revisited

It has been said that dead men tell no tales. Perhaps, but dead batteries may have something to say. So if you find a bug that has a dead battery, don't throw it away. It can be analyzed to determine when it died. An A-cell autopsy, as it were. As batteries discharge, the chemicals inside them change. Even after the battery is effectively dead, the chemicals continue to decompose, and analyzing them can narrow down the approximate date of the battery's demise. This can help determine the period of time that it was operating. Next, by checking your appointment schedules, you may be able to determine what information might have been compromised, and perhaps who might have wanted it.

Where Was That Li'l Critter Born?

In addition to what can be learned from the above, it may be possible to determine some other information about the bug. There are people who can analyze the critter and maybe tell you a lot about it. Such as who made it. (It is easy to tell a homemade transmitter from one made by AID or Lorraine or Cony.) One such person is Kevin Murray. Contact information is in Appendix C.

The range of the transmitter can also be estimated, narrowing down the location of the listening post, and the position of the antenna may help determine the direction. Then a careful analysis of the buildings within the probable area, and who inhabits them, could possibly lead you to the person(s) or the organization that installed the bug. Chances are somewhat remote, but possible through a careful process of elimination.

Surveillance transmitters are much like their biological counterparts—you never know where they might be hiding.

Transient Bugs

Bugs take up residence not only in homes and offices, but also in hotel and motel rooms. The li'l critters might be anywhere. Business persons often have secret meetings in hotel rooms, figuring that they are more secure there. Maybe. They also know to create some background noise just in case someone is listening. A TV set, for example. Or running water, which is quite effective unless the spy has equipment to filter part of it out.

OK, so you and your partners of the Red Widget Company check into the No-Tell Motel to discuss your plans to buy out the Blue Widget Company. Naturally, you don't want spies from Wexler's Widget Works to know about your plans. So you go to turn on the TV to cover your conversation, but it doesn't work. Hmmm. Pull the plug out and have a look. An old spy trick is to clip one of the prongs off to disable it. Another is to disconnect the antenna. If this has been done, get a little paranoid. Someone may be trying to eliminate sound that could interfere with their transmitter. You can make a search, but it would be simpler just to get a different room. Tell the manager that yours has "bugs" and things will happen much faster than most other complaints. And next time, don't use your office phone to make reservations. Call from a pay phone.

Defensive Measures

When the search is complete, there are a few things you can do to help prevent a spy from being able to install a surveillance device in the future. The main one is to prevent anyone from getting in by securing the area. While few homes or offices are completely burglar-proof, you can make it extremely difficult to get in. Alarm companies and locksmiths have products worth investing in, and crime prevention divisions of your local police department may have some useful information.

Closed circuit TV is becoming very reasonable in cost and might be considered.

The legal use of transmitters discussed earlier in this book is something to consider. There are ways to set up a transmitter so that if someone opens a door (or whatever) it will call your pager number. *Nuts & Volts* has a series of articles on this, and reprints are available.

Paladin Press has books on burglary and the prevention thereof. Also, you can paint over anything that could be opened to hide a bug, such as office equipment, wall plug covers, and whatever else. You can use clear fluid that glows under ultraviolet light, available at "spy" shops, or nail polish. If the "seal" is broken you will know at a glance that tampering has occurred.

You already know that in order for someone to install a tap within the premises, house, apartment, or office building, he has to get access to the line. So, the same preventive measures that would keep a burglar out apply here. Check to see if the connection box (66 block or NIC) is in a secured area. Locked and with limited access. If not, raise hell with the property owners.

You can also use the yellow and black alarm system explained in *The Phone Book*, or a perimeter alarm transmitter to alert you if someone enters the phone closet.

13

Frequently Asked Questions

SURVEILLANCE IN GENERAL

Q. Are there any kind of inside surveillance devices that cannot be detected—literally cannot be found?

A. No. Whatever type of device is used to intercept a conversation and send it to a remote location, whether it is via a radio signal, visible light, infrared, microwave, a pair of wires, or whatever, something at the other end has to be used to convert the signal back to audio. And whatever type of devices are used, if one person can get them, then so can another. Literally any eavesdropping device will be found in the physical search if it is done right, if it is thorough enough. The answer to the question is NO. Note, however, that we are talking about inside devices. There are directional microphones that the physical search may not locate if access to the area is not possible. Such as in a house across the street. There are also downline phone taps—taps located off the premises such as in an underground telco junction point or on a telephone pole. If you can access them, and know what to look for, you will find the tap. If not, then you won't.

Q. Aren't some devices more difficult to detect, to find, than others?

A. Indeed. The spread spectrum transmitters described in the text. The signal may be such that you have to be right on top of it to detect it. They can be located with a spectrum analyzer or one of the receivers from Marty Kaiser, but not by some "bug detectors." There is a subcarrier transmitter that is built into fluorescent light fixtures. As it is wired into a different circuit than the wall outlets, carrier current receiver (CCR) detectors won't hear it—they have to be connected to that circuit.

Again, they will be found in the physical search if it is thorough enough, but you have to draw the line somewhere. It isn't practical to tear the place apart, rip out the walls. That's where experience is important. Knowing what you are looking for— always having images in your mind, of knowing when something doesn't look right, seeing signs of tampering that the untrained eye would miss. This is something that is learned through experience but is also sometimes partly a gut feeling. In *The Axnan Attack*, by Glenn Whidden, the operative assigned to bug a conference room hollows out a block of wood and places the transmitter inside it. The trained eye would likely notice this, even though the wood was matched to the drawer in which it was placed, and remember about using an ultraviolet lamp, with which it would definitely be detected.

LASERS

Q. Is it true that lasers can be used to hear people talking through closed windows?

A. Yes, but . . . we built one of these systems in college lab. Here is how it (sometimes) works. A low-power laser is focused on the target window. The return (reflected) beam is received through a small telescope and focused onto a circuit that is connected to an audio amplifier. Sound in the target room causes the glass to vibrate. This makes a tiny change in the distance between the glass and the laser system, which is converted back to audio.

When a beam of light strikes a surface, it is reflected at the same angle at which it strikes. Shine a flashlight straight ahead into a mirror and the beam reflects back at you. Stand against a wall and hold it at a 45-degree angle and you'll see the beam reflected on the opposite wall. So, obviously the laser has to hit the wall dead on to be reflected back to its location to be captured by the receiving device. Otherwise, two locations would be required, which is not easy to set up and would impair its performance.

Also, they work only under optimum conditions. Fog, snow, rain, passing cars, aircraft, elevators in the building, etc. all interfere with the return signal. There are ways to clean up the audio—special filters, sophisticated equalizers, and digital signal processors—but they are of limited use. Lasers are also defeated easily. Close the drapes. Place a fan near the window. Tape a small transistor radio to the glass and treat the spies to a little Beethoven. If you want to see for yourself, you can buy one from Information Unlimited in Amherst, New Jersey. They have some really interesting stuff there, but don't expect the laser to work as well as you see on TV.

DATA ENCRYPTION

Q. I have heard that the government has computers that can instantly break any kind of encryption code. Is this true?

A. No. First of all, the term "break" is inaccurate, since it suggests that there is a way to decrypt any message encrypted with a particular cipher, even though the keys or passwords are different for different messages. This may have been true of the early letter substitution and exclusive-or (XOR) ciphers of the distant past and is definitely true of ciphers that have "trap doors" built in (such as Clipper undoubtedly did), but not with the extremely complex ciphers such as DES-X (Data Encryption Standard "X") and PGP (Pretty Good Privacy).

Attack Methods

Assuming that there are no such weaknesses, there are various ways to go about attempting to convert an encrypted message back to readable form, called plaintext. This is called an "attack." One of these attacks is called "brute force," in which every possible key is tried until the right one is found, or "derived." Whether or not this can be done depends on the length of the "password," called "keyspace" that the encryption algorithm (program) uses.

"If all the 260 million personal computers in the world were put to work on a single PGP-encrypted message, it would still take an estimated 12 million times the age of the universe, on average, to break a single message."

—William Crowell
Deputy Director, National Security Agency
March 20, 1997

The Data Encryption Standard, which has been around for about 25 years, encrypts text in blocks of 64 bits (8 characters) and uses a keyspace that is 56 bits long. Seven characters times 8 bits = 56. The remaining 8 bits are used for parity, an error checking mechanism. The DES has been successfully attacked by brute force in approximately 27 hours. This was recently announced by Computer Professionals for Social Responsibility (CPSR). And while no one is talking, it is probable that agencies such as the National Institute of Standards and Technology (NIST) and the Defense Intelligence Agency (DIA) can derive a DES key in real time. Ergo, the DES should not be trusted for anything important.

If the keyspace were to be increased by one, from 56 to 57, the amount of time required to try all keys doubles. So, CPSR might need 54 hours to decrypt such a message. Now, let's compare this with the International Data Encryption Algorithm (IDEA), which uses a keyspace of 128.

Suppose some agency, perhaps NIST, starts with a machine that can try 1 billion possible keys per second, and then parallels a million of these machines. All of them working together to derive the key of an IDEA-ciphered document. At that rate, not allowing for breakdowns, it would take about 10,782,897,524,600 years. Given that, on average, the right key would be found after trying about half the possible keys, it might be derived in about 5 trillion years. By then the information probably would no longer be of any importance to anyone.

Incidentally, if each machine (actually each individual circuit board) were 12 inches square, were stacked 10 high, and if a series of aisles were built so that repair technicians could access all of them in order to replace the chips that would be burning out on a frequent basis, such a machine would take up an area about the size of three football fields. If each circuit board used 40 watts, a nuclear generating plant would be required to power it. And another one to cool it.

So many personnel, technicians, programmers, etc. (and what with the way government does things, security and administrative people) would be required to operate and maintain it that there would be an unending line of trucks at the gate to deliver corned beef, rye bread, and Granny Goose potato(e) chips. The local Round Table Pizza would have to build a restaurant about the size of a blimp hangar, and the Jolt cola folks would probably build a bottling plant nearby.

All kidding aside, with the technology of today, there just isn't any way to brute force defeat the IDEA cipher. This is why the federal government is trying so hard to keep us from using it.

Another method of attack is known as "plaintext-cyphertext pairs," meaning that if both the encrypted message and the plaintext message are available, it is a little easier to find the key. Perhaps in a few thousand centuries . . .

"Why would anyone want to use encryption?" the government wants to know. After all, there's no reason to unless a person has something to hide. "They must be guilty of something . . ." Or, as Phil Zimmerman put it, "Why do people seal their mail in envelopes if they have nothing to hide? Why don't they just use postcards?"

Libraries

If an agency, such as the NSA, were to try and find a password, they might use something called libraries. These would consist of every single fact they could find about the person who encrypted the message(s). Not only personal information (past addresses, SSN, names of friends and neighbors past and present), but also the names of the streets in the city where they went to college, their instructors' names, terms used in the subjects they studied, names of well known authorities and experts, inventors, technical terms, anything to do with their hobbies and interests (photography, flying, sailing, etc.) literally anything they can find.

Q. What is PGP?

A. Pretty Good Privacy. Developed by Phil Zimmerman, it's a "public key" cipher, meaning that it has two keys. One you keep secret, and the other, the public key, you can make available to others. If someone wants to send you a secret message, they use your public key. Only your private key can decode it. Conversely, a message encrypted with your private key can be decoded only with your public key. PGP uses the IDEA described above to encrypt the message and then the RSA (Rivest-Shamir-Adleman) algorithm to encrypt the public and private keys. To learn more about PGP, connect to their Web site: www.pgp.com. A detailed, step-by-step explanation of how the RSA cipher works is on my Web site (www.fusionsites.com).

Again, if you use a long key, called a "keyspace," such as 1024 bits, PGP cannot be "broken" with the technology that exists today, and probably not for many years to come. Every time the NSA tries to catch up, an increase in the keyspace (length of password) of one bit would send them back to the drawing board. I am being uncharacteristically conservative here. People who know encryption far better than I suggest that "breaking" PGP would agree with the statement by the NSA official—that it would take longer than the age of the known universe. But again, it is really academic. After a few years, the information will probably be worthless.

Q. What is a safe password?

A. The best key for most ciphers would be a long string of random characters including letters (upper and lower case), digits, and punctuation marks. However, this is difficult to memorize and would require that it be written down, which is NOT a good idea.

With PGP, however, the program generates the keys for you based upon random numbers generated by moving the computer mouse around for a few seconds. Then you select a password with which to encrypt that set of keys. So, you can use a phrase that you make up, something that is easy to remember but means nothing to anyone else. "I'll never fOrget that SumMer eveNing in PawtuckEt in 1972," or "When I wAs a cHild, I had a pet aaRdvark named Andy," or a nonsensical phrase such as "Magic mageNta martians maKe marvelous mOOnbeam martinis." Or something like that. But be absolutely sure that you can remember it. If you lose or forget your password, anything encrypted with it is lost. THERE IS NOTHING ANYONE CAN DO TO GET IT BACK. IT IS GONE FOREVER.

Q. There are lots of ciphers, encryption programs, advertised on the Web. How do I know if I can trust them (i.e., that they are strong enough or that they do not have "trap doors"?

A. The first thing to ask is whether or not the source code (the program before it is compiled) is available. If it is not, the program should not be used. Period. This is not to say that such a cipher does, in fact, have a weakness, or that it is not strong enough; it is to say that without the source, you have no way of knowing what it is. For all you know, this could be a Trojan horse produced in the USSR. It has happened before. It could be produced by the NSA. Perhaps you are not aware of it, but the federal government has been spending a great deal of the taxpayers' money to prevent Americans from having secure encryption. Remember the Clipper chip the government tried to force upon us a few years ago?

Acceptance

No algorithm, no program, is accepted in the encryption community until it has withstood the test of time. Until it has been around for a few years and has been analyzed by the experts. And this, of course, requires that the source be available. The IDEA cipher took years to become accepted as secure against government attack. As stated elsewhere in this chapter—the purpose of encryption is to prevent unauthorized persons from accessing your information. Why compromise it with something you are not sure about? Especially when many of these proprietary ciphers sell for hundreds of dollars and one version of PGP is totally free.

TEMPEST & COMPANY

Q. Rumor is that it is possible to read everything on a computer screen from a great distance. Is this true?

A. Yes, but it is very complicated. TEMPEST stands for Transient Electromagnetic Pulse Emanation Standard, a system of measuring the "signals" that are "broadcast" from a computer.

The whole thing started many years ago when Dr. Win van Eck demonstrated that this technique actually works. Since then, this method of monitoring has been known simply as van Eck.

In the early days of computers, the monitors radiated signals that were so strong that they often interfered with TV reception, and through the use of a simple circuit, a TV could be modified to better receive these signals. Actually read everything on the screen. Much has changed since then, since the seventies, and today this is a much more difficult operation. The equipment is much more sophisticated—and very expensive. (See http://www.dynamic-sciences.com/tempest.html for details.)

There are many things that will help reduce the chances of someone's reading your screen from a distance. There are TEMPEST secure computer systems available in different levels of security. The most secure, Level III I think it is called, is restricted to government and some big businesses, but the others are available.

A little more practical and affordable is to place monitors in inside rooms or so that they don't face windows. Anything metal, especially if it is grounded, between the monitors and the outside world will help. Also, be aware that these signals can also be received from the main computer board, drives, modem, and printer. So, if you are concerned, make sure your computer case is grounded, that all cables are shielded. Or, there is a better way. Use a notebook computer. Far as I know, this—along with proper shielding—defeats van Eck monitoring completely. And large LCD monitors are appearing on the market that will accomplish the same thing. Expensive, but the price will start going down soon.

The Spy in the Keyboard

Here is an interesting gadget from C-Systems in the UK.

> The MicroSpy is a miniature computer in its own right. In operation it monitors the characters as they are typed in on the target computer keyboard and stores them in its own built-in memory. This special memory needs no batteries and can store more than 1000 keystrokes. Plug it in to the target system [in series with the keyboard cable] after it has been shut down for the day, and when the operator fires it up the next morning, it captures their user ID and password sequence, and if they log on to a network, that information is also stored.
>
> Then remove the device, connect it to your computer, run the included Windows software and the information is there on your screen.

Clever idea. Details and a picture of MicroSpy are at www.c-systems.co.uk/pc_keyboard_ computer_bug.htm

Video Surveillance

I know very little about video; just never was interested, so all I can say is what you probably already know. Video cameras are small, they're cheap, they can see in near total darkness, and they're available openly on the market to anyone who has the cash. Some are so small that they can be hidden virtually anywhere, including in baseball caps and even built into the frames of eyeglasses. They're everywhere. Banks and airports and convenience stores, hotel lobbies and train stations and retail shops. They're on the streets of most large cities, on the corners of buildings or inside fire extinguishers or disguised as power transformers on utility poles. They're in the workplace, the shop, the office, and elevators. Yes, in elevators, but you won't necessarily see them. If you are ever in the TransAmerica Pyramid in San Francisco, look at the light fixtures. One of them has a small, pencil-size hole. Behind it is a camera. Be careful what you scratch!

As to the law on video surveillance, I believe that it is not prohibited by Omnibus (or other laws) as is audio surveillance. I was a guest on Wisconsin Public Radio in January 1999 along with a video expert who explained that these cameras can be anywhere except where a person has a reasonable expectation of privacy. Rest rooms and dressing rooms, I guess this means. And, of course, they can not be installed anywhere without permission of the owner, so no one can legally set up video in your home, or on your property. As far as being off your property but focused on your property, this may be an invasion of privacy. Ask a lawyer.

Most of these cameras have recorders attached, but usually no one is actually watching the monitors; that would require personnel that few companies are willing to pay for. Some are changed when full, dated, and stored for a certain time. Others record on an endless loop. Then, if something happens, such as Granny Yokum trying to filch a flounder at Safeway, the tape will catch her in the act and can be used as evidence.

Here is something interesting to think about. If federal agents install a video camera in such a way that they can see people's faces, they could employ skilled lip readers to extract at least part of what is being said. Little doubt that they have been doing so for years. It may or may not be admissible as evidence, but this is one of those situations where they use surveillance on someone they suspect of something, and when they obtain sufficient information, then they apply for a surveillance warrant for a phone tap.

OTHER SURVEILLANCE METHODS

- *RF flooding.* This is a method of beaming a strong RF radio signal at a multi-line phone. Sometimes you will be able to hear conversations from other phones on the same system. This happened one time, completely by accident, when I was a technician at Countersurveillance Systems, Inc. We were testing some equipment, and I was on the phone when suddenly the boss' voice came blasting over the speakerphone from his office on the other end of the suite.

- *Microwave systems (both outside and inside) as used against the U.S. Embassy in Moscow some years ago.* Now, this is fascinating. One such system was the "Great Seal," which made headlines some years back and is described by Glenn Whidden in his book *The Russian Eavesdropping Threat in Late 1995* "The device was a passive cavity resonator that was concealed in a wooden plaque that has the seal of the United States carved on its front surface. It was presented as a gift from the Soviets to the American ambassador at Moscow. It was powered by a radio signal that was transmitted to it from a point outside the embassy and radiated a radio signal that carried room sounds from the Ambassador's office." Clever, these Russians.

- *Directional microphones, such as the type used by sportscasters at football games.* Very expensive (the good ones run a grand or more) but are available to anyone who has the cash. Sennheiser is supposed to be the best, and Shure and Electro-Voice are also excellent. However, don't expect the same results as you see on TV movies and PI shows; they have their limitations. Ambient (background) noise, sounds near the target, and wind noise all affect the signal. And the range (depending on conditions) is likely to be a few hundred feet at best. Not a "half mile" as some ads have claimed. Unless you like to listen to jackhammers . . .

The Final Word

OK, troops, having completed the text of *The Bug Book*, you have learned a fair amount about surveillance. You know the basics of planting a bug and tapping a phone.

You have also learned about countermeasures. How to obtain and use the equipment needed to find surveillance devices, and how to deal with them should they be found. This does not make you an expert. However, with what you now know and a few hundred dollars in equipment, you will be able to make it very, very difficult for anyone to bug your home or office.

Motive . . . method . . . opportunity.

There is a lot to absorb here, so may I suggest that you put it away for a week or so and then read it again. Refresh your memory. Study it until you can open to a page at random, zoom in on a paragraph, and understand what it says. Know what it means. When you are out dropping an RF transmitter in the mayor's office, you won't have time to be looking things up. And that's just as well, since I'd rather that you didn't have this book with you if you get caught.

But, perhaps this isn't good enough. Maybe you don't feel secure about taking on these countermeasures procedures. You want to call in a professional sweep team. If you do, please be careful about hiring someone who claims to be an "expert." There are maybe a dozen or so companies in the entire country that qualify as true professionals. That are equipped with $100,000 in equipment. The vast majority of self-proclaimed experts are not. They may be well meaning and honestly try to do a professional sweep, or they may be phonies who will lie to you and take your money.

How do you tell the difference? Get a copy of *The Phone Book* and read "On a Sweep." It is based upon my personal experiences as a sweep technician for one of the best teams in the country. It has photographs of the equipment as it is set up for a real sweep and a series of questions you can ask a prospective sweep team to determine whether or not they really know The Biz.

I hope that you never need the information in this, my last "spy" book. I hope you never have to deal personally with electronic surveillance. But most of all I hope, should this happen to you, that you are able to use the information in this book. That because of what I have written, you are able to successfully deal with the situation.

Damn right this book is necessary.

—M.L. Shannon
San Francisco
November 1999

APPENDIX

A

Some Classic Books

All of these titles are old, but they are excellent books. The technology may have changed in the years that followed their publication, but as you have read in the text, many of the principles of surveillance do not change. It is fascinating to read them and see how prophetic the authors were in predicting how the government would take control of technology, of communications, and how privacy was being systematically and insidiously taken away from We the People.

Many of these titles are available at public libraries in the larger cities, or through inter-library loan, and can be ordered through used bookstores.

Brown, Robert M. *The Electronic Invasion*. New York: J. F. Rider, 1967.

Callahan. *Cheesebox*. A long-out-of-print autobiography of one of the greatest wiremen of his time. Fascinating book! Worth searching for. (If you happen to have an extra copy, I would like to buy it.)

Campbell, Duncan. "Big Brother Is Listening: Phonetappers & The Security State." *New Statesman* (c. 1981): 2. Series title: NS report.

Carroll, John Millar. *The Third Listener: Personal Electronic Espionage*. New York: Dutton, 1969.

Cook, Earleen H. *Electronic Eavesdropping*. Monticello, Illinois: Vance Bibliographies, 1983. Series title: Public administration series bibliography P1262.

Cunningham, John E. *Security Electronics*. Indianapolis: H. W. Sams, 1970.

Dash, Samuel, Richard F. Schwartz, and Robert E. Knowlton. *The Eavesdroppers*. New York: Da Capo Press, 1971. Series title: Civil liberties in American history.

Fitzgerald, Patrick, and Mark Leopold. *Stranger on the Line: The*

Secret History of Phone Tapping. London: The Bodley Head, c. 1987.

LeMond, Alan, and Ron Fry. *No Place To Hide*. New York: St. Martin's Press, 1975.

Long, Edward V. *The Intruders: The Invasion of Privacy by Government and Industry*. New York: Praeger, 1967.

Pollock, David A. *Methods of Electronic Audio Surveillance*. Springfield, Illinois: Thomas, 1973. A classic!

Schwartz, Herman. *Taps, Bugs, and Fooling the People*. New York: The Field Foundation, 1977.

Spindel, Bernard B. *The Ominous Ear* (1st ed.). New York: Award House, 1968.

Wingfield, John. *Bugging: A Complete Survey of Electronic Surveillance Today*. London: R. Hale, 1984.

Glossary of Terms

AF—Audio frequency; the range of sound that people can hear. Usually stated as 20 to 20,000 cycles per second, though few people can hear this entire range.

AM—Amplitude modulation; a radio signal that uses changes in its amplitude (intensity) to carry intelligence or, in the case of television signals, lack of intelligence.

ANAC—Automatic Number Announcement Circuit, a service of telco that will identify the number of a line. Call the ANAC number and a computer voice will read back the number you are calling from.

ANTENNA (also sometimes called aerial)—Anything used to radiate a signal from a transmitter to increase its range, or to increase the ability of a receiver to hear a signal. An antenna may be as simple as a piece of wire or as complicated as a multi-element array.

BABY MONITOR—A type of monitoring device intended to listen for small children, such as babies in cribs. Most of them transmit on 49 MHz and can be used as bugs as they are easily available and cheap. Can be found in thrift stores and second-hand shops for about $5.

BIAS—A weak signal generated by some, but not all, tape recorders. It is used to align the small areas of magnetism on recording tape, called domains, previous to sound being recorded. This signal can sometimes be detected by special devices made for the purpose at very close distances. Such devices, however, will not work with the Nagra recorders that the feds use, because Nagra has a switch to disable the bias. Sneaky rascals, aren't they?

BODY WIRE—A small transmitter usually hidden under the clothing of a spy, agent, etc., used to transmit conversations without anyone but the other spies or agents knowing.

BRIDGING BOX (B-BOX)—A connection point, usually located on a street corner, where large feeder cables (F-1) connect to smaller distribution cables that lead to apartment and office buildings or private homes.

BRIDGING TAP—A physical electrical tap of a phone line.

BUG—Generic term that means different things to different people. A bug may be an RF room audio surveillance transmitter, a hidden microphone, or a phone transmitter.

BUG BURNER—A device used for destroying phone transmitters by sending an electrical charge through the phone lines. This should be used only by someone who knows what he is doing, since there is the danger of getting zapped. There is also the possibility of damaging the telco equipment, in which case Ma Bell's security people will track you down and kill you.

BUG DETECTOR—A device used for finding a bug by tuning in on the signal it transmits, which may be RF, visible, or infrared light. Not to be confused with Pest Control.

BURST—A transmitter that converts sound to digital form, stores it, and then transmits it in a short, high-speed burst. This makes it more difficult to find with electronic equipment because it does not transmit continuously.

CABLE—Any number of electrical conductors together, usually inside an insulating sheath. It may contain from one to thousands of individual wires.

CO—Central office of the telephone company.

CYCLE—The number of times per second an alternating or time-varying current changes direction, intensity, phase angle, etc. Also called Hertz.

DAMOCLES, SWORD OF—Something that materializes out of thin air and hangs over the head of a writer as a deadline approaches.

DBM—*Don't Bug Me*. No, this doesn't mean I don't want to hear from you; it is the title of my first Paladin book.

DECIBEL—A unit of measurement for the relative strength of an audio or RF signal. Named after Alexander Graham Bell.

DECOMPOSITION—The breakdown of chemicals in a battery. The chemicals can be analyzed to determine the approximate time period in which the battery was functional and a surveillance transmitter was operating. Part of "damage assessment," this can help determine what information may have been compromised.

DEMODULATOR—A circuit in some RF transmitter detectors that converts the signal into audio and feeds it into a speaker, so the operator can identify transmissions and eliminate false signals such as commercial radio and TV stations.

DISTRIBUTION CABLE—A smaller cable containing phone line pairs that branches off from the feeder cable at an appearance such as a B-box or junction point.

DTMF—Dual Tone-Multi Frequency. "Touch-Tones," i.e., the audio tones used by telephones to dial numbers and other functions. The standard phone has only 12; 0 through 9, *, and #. Some phones, such as used by the military, also have A, B, C, and D, which have special functions.

DROP—Make a drop; plant, hide, or install a bug or listening device. Also, what you do when someone catches you installing a bug and shoots you with a .357 magnum.

EFFICIENCY—The ratio of the output power to consumed power. A circuit—an amplifier, for example—that puts out 100 watts and draws 200 watts from the power line, is 50-percent efficient.

ELEMENTS (antennas)—The parts of a beam, or yagi antenna: directors—the parts that concentrate the signal in one general direction; driven element—the part that is connected to the transmitter; reflector—the part that reflects to signal from the driven element in the desired direction, where it is concentrated by the directors.

FEEDER—A cable (typically 1,200 pairs) that goes from the telco CO to an appearance such as a B-box or junction point. From there, it splits into distribution cables that go to homes, offices, and apartments.

FIELD STRENGTH METER—An electronic device that can detect the RF signals from transmitters, bugs, etc. It can be as simple as a small meter with a diode across it and a length of wire for an antenna, or as sophisticated and expensive as the models made by Simpson, and others, used to detect leakage in cable TV lines.

FM—Frequency modulation. An RF signal that uses changes in frequency to carry intelligence as opposed to changes in (AM) amplitude. Also, the FM commercial broadcasting band, named so because the stations operating there use frequency modulation transmitters.

FREQUENCY COUNTER—A device that measures and displays the frequency of a radio transmitter.

FREQUENCY SPECTRUM—The area of the total electromagnetic spectrum in which radio waves operate. The spectrum is from zero cycles per second, at the bottom of the subaudio band, to cosmic rays. The part that is used for radio signals is from about 20,000 cycles, the top end of the audio band, up to where infrared light begins, at about 300 GHz.

FULL DUPLEX—A radio or radiotelephone system where both parties can talk and listen at the same time. Cellular telephone, for example, is full duplex.

FULL QUIETING—The condition where the signal from a transmitter is heard by the receiver at such a level of signal strength that it blocks out all background noise.

HACKER—Many definitions apply. In the true sense of the term, it refers to those who have an interest in understanding the complexities of things: the telco equipment, the Internet, computers in general, and so on. True hackers are not malevolent and do not destroy facilities or steal information with the intent of depriving the owner of the use and enjoyment of such information and realize no monetary gain from their explorations.

HARMONIC—A multiple of an RF signal. For example, a transmitter with a signal on 10 MHz would also transmit on 20, 30, 40 MHz, etc. These harmonics are usually suppressed in the transmitter, are very weak, and don't radiate very far. But a "dirty" bug may have strong harmonics, and they might be seen on a TV screen in the bugged area.

HEAT COIL—An overload protection device installed on the telco frames to prevent damage to the equipment from lightning striking the lines, idiots plugging the phone line into a wall socket (it happens!), or inexperienced personnel using a bug burner.

INTELLIGENCE—Fancy term for information, often used by people who are trying to impress others.

INTERCEPT—To overhear, in any of several ways, conversations without the subjects' being aware that it is being done.

JUNCTION POINT—An underground room, usually entered through a manhole cover at a street intersection where telco lines are accessed. It is used for splicing, repairs, creating traffic jams, and sometimes wiretapping. See also Bridging Box.

LISTEN-DOWN AMPLIFIER—An audio amplifier connected to a phone line. It allows the user to hear anything on the line without seizing it (the phone is still on hook).

LISTENING POST—Any place used to intercept information from a listening device. It can be on the premises, in another apartment or office, in a van parked nearby, or miles away.

LOCAL LOOP—The "loop" of wire that connects an individual telephone to the central office.

LOOP EXTENDER—A pair of wires (a phone line) used to connect a tapped line to the listening post. Law enforcement have these lines leased from the telco to use for telephone surveillance.

MULTIPATH DISTORTION—Receiving a signal directly from the transmitter as well as indirectly because the signal has been reflected from something in the signal path. Bounced off a mountain, office building, passing aircraft, whatever. The reflected signal arrives at the receiver a few microseconds behind the direct signal because of the increased distance between the two. On a TV set, this is seen as "ghosts."

OFF HOOK—The condition of a telephone line when it is in use. The line voltage, normally about 48 volts, drops to about 10 or 12. If you had a meter connected to the line, you would see the change in reading and, if an extension were picked up, it would drop an additional two to four volts.

PENETRATE—To physically enter the target area to place (drop) a listening device.

PHANTOM LINE—Also phantom pair; the process of using one wire from one cable or pair, and one from a different cable or pair, in wiretapping.

PHONE TAP—Connection of a wire, bug, microphone to a phone line, or placement of a coil of wire on or near a line to intercept conversations.

PHYSICAL SEARCH—The process of physically searching for listening devices.

POINT OF DEMARCATION—The point at which phone lines become the property (and responsibility) of the telco. Usually beyond the 66 block in an office building, or SPSP in a private residence.

POLARIZATION—The way an antenna is positioned in relation to the earth, either vertically or horizontally. For best reception, the antenna should have the same polarization as the transmitter.

PRE-EMPHASIS—(simplified explanation) A circuit in a transmitter that amplifies (emphasizes) the high audio frequencies. In a surveillance transmitter, it makes the intercepted conversation easier to understand.

PROFILE—The composite information one might obtain about a person who has installed a listening device, based on various facts such as the type of bug used, where it was placed, and so on.

REMOBS—Remote observation. This is a function of the telco ESS with which authorized personnel can monitor any phone line in the system, possibly excluding lines that have special protection such as provisioning. See *The Phone Book*.

REMOTE CONTROL BUG—An RF surveillance transmitter that can be turned off and on from a distant location. This makes it harder to find and conserves battery life. If someone is listening and hears the sounds of a search, or a boring conversation, they can turn it off until later.

REPEATER—A system that uses a larger, higher power transmitter to relay a signal from a smaller one, thereby increasing the range. Also, an amplifier placed along telephone lines to boost the signal when it gets weak.

RF—Radio frequency; an area of the electromagnetic spectrum used for radio and TV transmissions.

SAS—Surveillance Administration System. A function of the telco with which the feds can tap any line they want to without having to set up a bridging tap and loop extender at the switch or other appearance. It has the FBI's Good Housebreaking Seal of Approval.

The following document on SAS is available on one of several CDs from Bellcore at http://telecom-info.bellcore.com/site-cgi/ido/index.html. The cost is $613.

Abstract—this document provides generic requirements for a wireline switching feature in support of the Communications Assistance for Law Enforcement Act (CALEA) of 1994.

CALEA requires that a service provider be able to support lawful surveillance of the traffic in its network. Lawful surveillance is the process of identifying a targeted set of traffic and delivering data about the traffic, and possibly the content of the traffic itself, to a remote law enforcement agency in real time. The document consists of Access, Delivery, Service Behavior, Administration, Security, and Maintenance sections. The document contains information from standards work on CALEA (e.g., TIA/T1 J-STD-025, ANSI, T1.260) but specifies additional requirements necessary for the implementation of this feature. A companion document is GR-2975-CORE, Surveillance Administration System (SAS). There are several other, similar, documents for sale. They will explain how SAS works, but they will not tell you specifically how to use it or provide access codes to the switch.

SEIZING—Answering a phone line; picking up the receiver; causing it to be busy or in use, in an off hook condition.

SENSITIVITY—The ability of a radio to hear a signal. This is expressed in microvolts or microvolts per meter. It is the amount of voltage required on an antenna one meter in length to produce a certain signal-to-noise ratio.

SIXTY-SIX BLOCK—Generic name for the electrical panel used for connecting incoming phone lines to the cables leading to offices or apartments. Named for having a capacity of 66 lines, it also applies to other size panels.

SPSP—Single pair station protector; a small metal or plastic can or box used to connect a single phone line, used in private homes and small apartment buildings. It contains fuses to protect the line and telco equipment from voltage surges, and people who use a bug burner without reading the instructions first.

SWITCH—The telco computer.

TARGET—The area to be bugged or line to be tapped.

TELEVISION—An IQ-lowering device.

TORX SCREW—A special fastener used to lock some telco connection blocks, TV cable converters, etc. It can be removed with a Torx wrench. What else?

TSCM—Technical Surveillance Countermeasures. The art and science of defense against surveillance through physical and electronic searching for eavesdropping devices, preventive measures, and education.

C

Places to Get Things

Alltronics

A legal use for transmitters, described in the text, is a signaling device. One such product is the Signamail, available from Alltronics for $14.95. There is a tilt switch in the transmitter, originally intended to be fitted to a mailbox, so whenever it is opened it sends a signal to the nearby receiver causing it to turn on a light and play a built-in tune. ("Return to Sender" . . . ?) However, the switch can be modified easily for use on a door, gate, or whatever. Also, the receiver can be adapted to turn on a siren or other alarm device. Maybe a special effects recording of an AK-47, which should effectively scare burglars away.

The transmitter could be modified with a "Panic Alarm" switch and be carried by children who are playing near their home or perhaps set up in the yard somewhere. Their parents, hearing the alarm, would be able to rush out and see what was the problem. If one of these devices were to save the life of one little kid . . .

They also have several transmitter kits, from 7 to $13, which are probably single transistor modulated oscillators, a $50 voice changer that I suspect is as good as some others costing twice as much, Torx wrench sets for opening things that you probably shouldn't, and a field strength meter for detecting hidden transmitters. This is a kit, requiring a minimum of electronics experience, and sells for $29. Haven't tried it, but looks like a fairly good product.

Alltronics
2300 Zanker Road
San Jose, CA 95131
www.alltronics.com

C-Systems

Located in the UK, this is a source of lots of fascinating gadgets. A wide range of crystal-controlled surveillance transmitters in the $300 to $400 range, a Global Positioning System tracking unit, countermeasures receivers, and other goodies described in the text. Note that it is unlawful to import many surveillance devices, so don't be surprised if U.S. Customs seizes your goodies.

C-Systems
P.O. Box 91
Leeds, LS12 4XR
England
Email: c-systems@zetnet.co.uk

Electronic Rainbow

This company sells mostly kits, all of which are inexpensive. For example, the FMST-100 stereo transmitter, tunable from 76 to 108 MHz FM. Specs say the range is about 200 feet, which in many cases is enough, but can presumably be increased by using a different antenna. As it is stereo, you can use two microphones in different locations, using the balance control of the receiver to select which one you want to listen to. Operates on a 9-volt battery operation. Cabinet size is 1.5 by 2.5 by 3 inches. Antenna is not included. Price, including case, is $38.90 plus shipping and handling.

Electronic Rainbow
6227 Coffman Road
Indianapolis, IN 46268
www.rainbowkits.com/index.html

Future Scanning Systems

FSS produces the excellent computer-aided scanning software I reviewed in the text. Far as I am concerned, it is the best for countersurveillance—and at least as good as any other for general scanning use. There is a demo available, and if you register it you get an excellent manual on disk, as well as tons of preprogrammed frequency lists. Cost is only $45, which is one helluva deal. It really is that good.

Future Scanning Systems
6105 SE Nowata Road #6
Bartlesville, OK 74006
www.futurescanning.com

Grove Enterprises

Grove is an old company—it has been around for many years and has a reputation for excellence in products and customer service. Grove has a large selection of scanners, communications receivers, and accessories, books, antennas—everything to do with monitoring. It also has sales, specials on used gear (with warranties), and interesting things to read online as well as in its magazine. Prices are competitive. Call or write for Grove's catalogue and see for yourself.

Grove Enterprises
P.O. Box 98
7540 Hwy. 64 West
Brasstown, NC 28902
www.grove.net

Ham Radio Outlet (HRO)

HRO specializes in ham radio equipment but has a large selection of communications receivers, such as ICOM and Yaesu and other brands. HRO is an old, established company; it has been around for many years. Its staff is experienced and professional. Whatever you need, HRO has it, or can tell you where to get it. HRO has stores in many locations, including

Ham Radio Outlet
510 Lawrence Expressway
Sunnyvale, CA 95051
www.hamstore.com

JDR Microdevices

This is an interesting store that I used to go to before I moved away from the choking smog of the South Bay to the clear air of San Francisco. Among its many products are a TV transmitter kit for $22 that can also be used to send audio on TV channels 2 to 6 (54–78 MHz). Range is listed at 300 feet, but as we know, this depends upon many things.

JDR has a DTMF decoder kit with large red LED readout. It can store 256 digits even when power is disconnected. Sells for $99, but you also need the power adapter ($7.99) and the optional plastic case ($14). If this is more than you want to spend, there's a kit without the LED display for $29.99, and the individual LEDs, which can be connected easily, are available for a few bucks at surplus stores.

There's also a cheapo frequency inversion voice scrambler for $39.99 and a light beam communicator that transmits voice over an infrared beam to a distance of about 30 feet and works with microphone input for $21.99. And should you believe someone has installed one of these in your bedroom, JDR also has an IR detector that plugs into the parallel port of a computer. The ad says it will "see every code pattern from any remote control or other infrared device." Nifty. And there's an FM wireless microphone kit for $34.99 that claims an unrealistic range of "up to 1 mile." Last but not least, something very interesting. Most of the "bugs" available are very low power; ergo they have very limited range (all things considered). This device is a linear amplifier that will boost the power to a full watt—a very large bug. Range is listed at 100 KHz to 1 GHz. Great for those long-range applications, but, of course, it has to have constant power.

JDR Microdevices
1850 S. 10th St.
San Jose, CA 95112
www.jdr.com

Marlin Jones

This company has all kinds of interesting gadgets that can be useful in surveillance and countermeasures. There are several room audio transmitters, one of which has three transistors and costs only $13; phone line transmitters; dropout relays; DTMF decoders; relay systems that can be used to turn things off and on by calling your number and entering a "secret" code; infrared motion detectors and transmitters; remote phone ringers so you can isolate your phone from the line and not miss incoming calls; caller ID boxes; telephone line cassette recorders; and video cameras/transmitters. The audio and phone transmitters are in kit form, so some assembly experience is recommended to build them. For hobbyists and technicians, there is also a large selection of tools, parts, power supplies, cabinets, and enclosures. The illustrated 135-page catalogue is free.

Marlin Jones & Associates, Inc.
P.O. Box 12685
Lake Park, FL 33403
www.mpja.com

Marty Kaiser Electronics

Marty makes a wide range of equipment, including the countermeasures receivers pictured here, and products for the detection of explosives (bombs) for use by law enforcement agencies. Everything Kaiser makes is high quality.

Martin Kaiser Electronics
P.O. Box 171
Cockeysville, MD 21030
www.martykaiser.com

MoTron

MoTron is the manufacturer of a number of products, including the TDD-8X, which decodes all 16 Touch-Tone digits: 0 to 9, # and *, A, B, C, and D (the last four are used in some military phone systems). It comes assembled and tested, with a large, bright red LED 8-digit display, and scroll right and left buttons to view its 104-digit memory. It also has a clear button to erase what is stored. To use it, plug in a 12-volt adapter and a cable from the decoder's audio jack to a scanner or tape recorder. You can also dump its memory to a computer through a serial port to maintain a record of all calls placed from the line it is connected to. MoTron makes excellent equipment and is highly recommended.

MoTron Electronics
310 Garfield St., Suite 4
Eugene, OR 97402
www.motron.com

Kevin Murray

Murray Associates
POB 668
Oldwick, NJ 08858
www,spybusters.com

Nuts & Volts

Nuts & Volts is a fascinating publication of how-to articles, columns, and advertisements (personal and commercial) for just about anything to do with electronics, computers, and communications. Back issues, which are available, have many articles on defense against surveillance and home security in general. An excellent publication. I subscribe to it.

Nuts & Volts
T & L Publications
430 Princeland Court
Corona, CA 91719
www.nutsvolts.com

Optoelectronics

I cannot say enough about Opto. It has portable, pocket-size frequency counters from $139 to $349 and bench models to about $700. They also have preselectors, which amplify signals to increase the sensitivity and effectiveness of receivers, and a computer control package for the PRO-2006 scanner. Two of their products that are of special interest in surveillance are the Scout, reviewed in *The Phone Book*, and the Xplorer, pictured in this book. Opto makes only very high-quality equipment that I recommend without hesitation to anyone.

Optoelectronics
5821 NE 14th Ave.
Ft. Lauderdale, FL 33334
www.optoelectronics.com

PhonicEar

This company makes a wide range of products for hearing-impaired persons, many of which are quickly and easily adaptable as surveillance devices. They operate on both infrared and FM, including the 72 MHz area mentioned in the text. Naturally, their products were not intended for electronic spying, but if thy neighbor buys one to bug thee, he probably will not announce his intentions to the dealer.

Phonic Ear, Inc.
3880 Cypress Drive
Petaluma, CA 94954
www.phonicear.com

Scanstar

This is one of the better programs with which a computer can control a scanner or communications receiver. It is powerful, versatile, and has many features and unlimited channels and banks. It takes some time to learn well but is worth the effort. Check out its Web site for details and a free demo. Available from Signal Intelligence:

www.scanstar.com/

Sheffield Electronics

The owner, Winston Arrington, is the author of *Now Hear This*, mentioned in the text. Sheffield also makes the Model 262 countermeasures receiver. A photo and specs are in *The Phone Book*.

Sheffield Electronics
P.O. Box 377940
Chicago, IL 60637

Shomer-Tec

Located in the Seattle area, this old, established company sells all manner of equipment. Much of it is for (and sales may be restricted to) law enforcement, such as batons, handcuffs (in case you catch a bugger in the basement), flashlights, Mace, laser sights, smoke bombs, perimeter security systems, lock picks, and much more. As to surveillance and countermeasures, it has an extensive line of telephone products: a device that can break into answering machines by remote control and another that will protect against such a device; long-play cassette recorders; Caller ID blockers; call screening devices; and the list goes on. Shomer-Tec is an excellent company. I have dealt with Shomer for many years, and I highly recommend them. They are honest people who stand behind what they sell. Call to get their large, informative catalog.

Shomer-Tec
P.O. Box 28070
Bellingham, WA 98228

V-ONE Corporation

The secure paging system mentioned in the text is available from:

V-ONE Corporation
20250 Century Blvd., Ste. 300
Germantown, MD 20874
www.v-one.com/vpn_news/1997/news-08-28-97.htm

APPENDIX

D

Frequencies Used in Surveillance

*P*ractically anything that transmits can be used, with varying degrees of success, for surveillance, and there is little doubt that most of them already have. Similar lists are published now and then in *Monitoring Times* and *Popular Communications*, both of which are excellent magazines and worth subscribing to. So, should you believe that thy neighbor has bugged thee, here are some bands to search.

CORDLESS PHONE BUGS

The new UHF cordless phone frequencies are between 902 and 928 MHz. Different manufacturers use different frequency groups within this 26 MHz wide band.

The Tropez DX900 has 20 channels spaced 100 KHz apart, with the handset 20 MHz higher than the base:

> Base   905.6 to 907.5
> Hset   925.5 to 927.4

So the base frequencies would be 905.6, 905.7, etc.

The AT&T model 9120:

> Base   902.0 to 905.0
> Hset   925.0 to 928.0

The Otron model CP-1000:

> Base   902.1 to 903.9
> Hset   926.1 to 927.9

The Panasonic model KX-T9000:

Ch	Base	Hset	Ch	Base	Hset
01	902.100	926.100	31	903.000	927.000
02	902.130	926.130	32	903.030	927.030
03	902.160	926.160	33	903.060	927.060
04	902.190	926.190	34	903.090	927.090
05	902.220	926.220	35	903.120	927.120
06	902.250	926.250	36	903.150	927.150
07	902.280	926.280	37	903.180	927.180
08	902.310	926.310	38	903.210	927.210
09	902.340	926.340	39	903.240	927.240
10	902.370	926.370	40	903.270	927.270
11	902.400	926.400	41	903.300	927.300
12	902.430	926.430	42	903.330	927.330
13	902.460	926.460	43	903.360	927.360
14	902.490	926.490	44	903.390	927.390
15	902.520	926.520	45	903.420	927.420
16	902.550	926.550	46	903.450	927.450
17	902.580	926.580	47	903.480	927.480
18	902.610	926.610	48	903.510	927.510
19	902.640	926.640	49	903.540	927.540
20	902.670	926.670	50	903.570	927.570
21	902.700	926.700	51	903.600	927.600
22	902.730	926.730	52	903.630	927.630
23	902.760	926.760	53	903.660	927.660
24	902.790	926.790	54	903.690	927.690
25	902.820	926.820	55	903.720	927.720
26	902.850	926.850	56	903.750	927.750
27	902.880	926.880	57	903.780	927.780
28	902.910	926.910	58	903.810	927.810
29	902.940	926.940	59	903.840	927.840
30	902.970	926.970	—	———	———

Present VHF cordless frequencies:

Ch	Base	Hset
01	46.610	49.670
02	46.630	49.845
03	46.670	49.860
04	46.710	49.770
05	46.730	49.875
06	46.770	49.830
07	46.830	49.890
08	46.870	49.930
09	46.930	49.990
10	46.970	49.970

Proposed new VHF cordless phone frequencies. These new channels are supposed to be made available in the VHF band:

Ch	Base	Hset
01	43.72	48.76
02	43.74	48.84
03	43.82	48.86
04	43.84	48.92
05	43.92	49.02
06	43.96	49.08
07	44.12	49.10
08	44.16	49.16
09	44.18	49.20
10	44.20	49.24
11	44.32	49.28
12	44.36	49.36
13	44.40	49.40
14	44.46	49.46

BABY BUGS

Wireless "Baby Monitors" such as made by Gerry or Fisher-Price make excellent bugs, as described in the text. Since they have a built-in power supply, they were made for continuous duty. They use some of the cordless phone frequencies:

Ch	Freq
A	49.830
B	49.845
C	49.860
D	49.875
E	49.890

DEAF BUGS

Transmitter/receiver sets made as auditory aids for the hearing impaired are used in lecture halls and movie theaters. They can be improvised to work as room audio or phone line transmitters. Older models will use two bands:

72.025 to 72.975 Transmit
75.475 to 75.975 Receive

The newer ones may operate on the 906 to 928 band.

INSIDIOUS BUGS

A technique for making a phone transmitter difficult to detect with countersurveillance equipment is to set the frequency very close to another signal that is very strong. This is called snuggling. You remembered that, didn't you? The audio portion of a TV signal is one likely place for this type of bug.

Channel 02: 059.75
Channel 03: 065.75
Channel 04: 071.75
Channel 05: 081.75
Channel 06: 087.75
Channel 07: 179.75
Channel 08: 185.75
Channel 09: 191.75
Channel 10: 197.75
Channel 11: 203.75
Channel 12: 209.75
Channel 13: 215.75

CELEBRITY BUGS

Some commercial wireless microphone frequencies to try are 169.45, 169.505, 170.245, 170.045, 171.105, 171.845, and 171.905, but they may be anywhere in this area. The UHF band is also now being used for wireless microphones. There are 50 channels between 794 and 806 MHz; unused TV channels 68 and 69.

HIGH FAT BUGS

The drive-up windows at fast food restaurants use wireless headsets that allow their employees to communicate with customers. These headsets are available commercially and, as far as I know, to anyone who wants to purchase them. Ergo, they could be used as bugs. The following, which are subject to change, are some of the frequencies upon which they operate:

McDonalds
35.020 and 154.600
30.840 and 154.570
33.140 and 151.895

Burgerville
30.840 and 154.570

Burger King
467.825 and 457.600

Hardees
30.84 and 154.57

TOY BUGS

Inexpensive "walkie-talkies" (no one really calls them that any more) available from Toys R Us may be used as surveillance transmitters. I have never had one, so I don't know if they will transmit continuously. Read the directions before you buy one.

916.8750	Ch. 01 (Calling Frequency)
915.8625	Ch. 02
915.0000	Ch. 03
914.0875	Ch. 04
913.3375	Ch. 05
912.0000	Ch. 06
910.9125	Ch. 07
910.2375	Ch. 08
909.3375	Ch. 09
908.5000	Ch. 10
907.6625	Ch. 11
907.0000	Ch. 12
906.3375	Ch. 13
905.6625	Ch. 14
904.5000	Ch. 15
904.0000	Ch. 16
903.4875	Ch. 17
903.0000	Ch. 18
902.5000	Ch. 19

E

A Physical Search Checklist

As I mentioned some pages back, there are many self-proclaimed countermeasures experts who are not. So, if you do decide to call in someone to do the sweep, please: first get a copy of *The Phone Book* and read Part 6: "On a Sweep." The forty bucks the book costs might save you thousands of dollars, but even more important, it may save you from being sold a false sense of security by someone who deals in such a vacuous commodity.

"On a Sweep" has a list of equipment that a professional will have, including at least one good countermeasures receiver, a spectrum analyzer, a carrier current receiver, a sophisticated antenna setup, a telephone analyzer, and many other devices and tools. It has a detailed 16-page description of an actual sweep (including photographs of the equipment as it is set up) from the initial contact to the follow-up and written report and everything in between. It also has a list of pertinent questions to ask someone who claims to be a countermeasures technician, and if he cannot come up with the answers the chapter provides, you will know not to hire him.

If you are unable to find, or afford, a suitable search team, please don't feel as if you are on your own. Get the best equipment you can obtain, use it to the best of your ability, and then do the physical search yourself, knowing that with what you have learned so far, you will be able to do just as good a job on the physical as someone with experience if you take the time to do it right.

You have that knowledge with you, and you always have this book as well as *The Phone Book* to fall back on and refer to, to be sure you aren't overlooking anything.

Have a look at the checklist of possible hiding places for the li'l critters that may compromise your privacy and security. It is not a complete list; there are too many hiding places to include here. It is intended to be a sample to generally cover most homes and offices and to get you to think. The list is sorted by area: basement, bathroom, bedroom, car, closet, garage, living room, office, perimeter, and utility room,and has places that are most likely to be found in such areas. It starts off with "All," listing things that might be found in any area. It is double spaced so that in your search you can add things you see that should be investigated. The brackets on the left are, of course, to check off everything that has been eliminated.

A good way to start is to make up a diagram of the rooms to be searched. Map everything that could possibly hide a listening device. This makes it easier to keep track of what has and has not been checked out. If you have never done this before, then you don't realize how easy it is to overlook something, and that is a no-no. Another useful technique is to stick those little round adhesive labels (Avery or Dennison) on things that have been centrally scrutinized. You might also draw a series of grid lines and check each section separately, crossing it off when complete.

While you are making up the drawing or looking over the checklist, I hope you remember from the text that one of the many secrets to successful surveillance is to think in ways that are unexpected, to do things that are unconventional, to flaunt tradition and use imagination. To install listening devices in places that are unexpected. You might wonder about some of the items listed here, so consider: it would be most unusual for anyone to have a ping pong table in the bathroom (even in San Francisco) but many people have a bathroom in the basement rec room.

A surveillance transmitter was once concealed inside a large, very large, black candle that was delivered to the victim as a gift. It was, of course, battery powered, and so had a limited life, but it was able to accomplish the objective. Would you have thought to check out a candle?

A transmitter was once hidden inside a hollowed out two by four, which was placed on the floor of a garage. Would you have thought about this?

You might not think that the kitchen is a logical place for a bug, but suppose the targets are the CEO and CFO of Wexlers Widget Works, and that both are gourmet cooks who get together now and then to whip up something exotic as they discuss a coming hostile takeover of Wortley's Wabbit Warrens. Not such a bad idea after all.

Remember that a bug can be hidden inside of, or

disguised as, virtually anything. So, reduce everything you see to its smallest part, until you know that nothing is there that shouldn't be. However, approach this with a little common sense. There are limits to your search. There has to be, unless you want to punch huge holes in the walls and tear up the floor, and effectively reduce the area being searched to rubble. If you really want to see what, if anything, is behind the walls, there are fiberoptic "bore-scope" devices available.

Consider things that cannot be moved without a great deal of effort, and which obviously have not been moved in a long time. Suppose you have a huge solid oak case full of books. It's been there for years and if someone were to move it to install a surveillance device, you would see indentations on the carpet or scratches on a hardwood floor. So, there is little need to move it to look under it, but of course you will want to look behind it and inside it. Including inside each and every book. You never know what you might find in old books. Like the fifty bucks you stashed several years ago and then forgot about.

A final thought: the physical search takes total concentration. Don't let anything interfere. No interruptions, no people in the area that are not part of the search team, no radios playing or phones ringing. Do the search in shifts as is necessary, because it takes your total attention and can be exhausting. Take a break when you start to feel fatigued. Otherwise, you are likely to get careless and make mistakes. And you know what the consequences can be. If you are not prepared and equipped to do the search thoroughly, then there is no point in doing it at all.

Surveillance happens. And it could happen to . . .

[ ] All radiator covers
[ ]
[ ] All phones, inside
[ ]
[ ] All phones, receiver
[ ]
[ ] All appliances
[ ]
[ ] All wigs, on stand
[ ]
[ ] All windows, curtains, behind
[ ]
[ ] All windows, rods, curtains
[ ]
[ ] All doors, maintenance access
[ ]
[ ] All mirrors, behind
[ ]
[ ] All clothes in closet
[ ]
[ ] All sculptures
[ ]
[ ] All flower pots
[ ]
[ ] All wall lamps
[ ]
[ ] All canes, hollowed out
[ ]
[ ] All light fixtures, ceiling
[ ]
[ ] All door frames, behind
[ ]
[ ] All picture frames, behind
[ ]
[ ] All picture frames, inside
[ ]
[ ] All carpet, under edges
[ ]
[ ] All phone jacks, behind
[ ]
[ ] All newspapers, rolled up
[ ]
[ ] All attaché cases, forgotten
[ ]
[ ] All ceilings, drop or false
[ ]
[ ] All doorknobs
[ ]
[ ] All doors, hollow
[ ]

[ ] All drawers, false bottom
[ ]
[ ] All mirrors, inside frame
[ ]
[ ] All cameras
[ ]
[ ] All cameras, film cans
[ ]
[ ] All cameras, modified lenses
[ ]
[ ] All bulbs, light, burned out
[ ]
[ ] All phones, wall, behind
[ ]
[ ] All windows, hung outside
[ ]
[ ] All furniture, upholstered
[ ]
[ ] All furniture, hollow legs
[ ]
[ ] All candles, hollowed out
[ ]
[ ] All flashlights
[ ]
[ ] All clocks, wall
[ ]
[ ] All floor, loose tiles or panels
[ ]
[ ] All musical instruments
[ ]
[ ] All stuffed animals
[ ]
[ ] All windows, outside of
[ ]
[ ] All baseboards, behind
[ ]
[ ] All boxes, storage
[ ]
[ ] All air conditioners, in, under
[ ]
[ ] All batteries, hollowed out
[ ]
[ ] All smoke detectors
[ ]
[ ] All alarm bells, burglar
[ ]
[ ] All registers, hot air
[ ]
[ ] All registers, air conditioning
[ ]

[ ] All vents, air
[ ]
[ ] All phone jacks, inside
[ ]
[ ] All waste baskets, under
[ ]
[ ] All windows, blocks, wood
[ ]
[ ] All wall plugs
[ ]
[ ] All wall switches
[ ]
[ ] All magazines, rolled up
[ ]
[ ] All paneling, behind loose
[ ]
[ ] All plug-in adapters for radios
[ ]
[ ] All boxes, phone, old
[ ]
[ ] All doors, frame, top
[ ]
[ ] All ash trays, hollow
[ ]
[ ] All drawers, under
[ ]
[ ] All metal surfaces, magnet box
[ ]
[ ] All heaters, electric
[ ]
[ ] All rods, curtain, hollow
[ ]
[ ] All waste baskets, false bottom
[ ]
[ ] All clocks, desk
[ ]
[ ] All dividers, room
[ ]
[ ] All heaters, gas or oil
[ ]
[ ] All alarm bells, fire
[ ]
[ ] Basement workbench, under
[ ]
[ ] Basement washer, drier, under
[ ]
[ ] Basement plumbing, unused
[ ]
[ ] Basement storage trunks
[ ]

[ ] Basement ping-pong table, under
[ ]
[ ] Bath dispenser, Kleenex
[ ]
[ ] Bath toilet, paper roll
[ ]
[ ] Bath medicine cabinet, in and behind
[ ]
[ ] Bath razor, electric
[ ]
[ ] Bath toilet, float
[ ]
[ ] Bath soap, hollowed out
[ ]
[ ] Bath jars, cosmetic
[ ]
[ ] Bath tube, toothpaste
[ ]
[ ] Bath toilet, tank
[ ]
[ ] Bath toilet, behind
[ ]
[ ] Bath deodorizers
[ ]
[ ] Bath drug bottles
[ ]
[ ] Bath mirror, make-up
[ ]
[ ] Bed jewelry boxes
[ ]
[ ] Bed box, hat
[ ]
[ ] Bed mattresses
[ ]
[ ] Bed dolls, hollowed out
[ ]
[ ] Car taillight
[ ]
[ ] Car bumper beeper
[ ]
[ ] Car mirror, rear view
[ ]
[ ] Car upholstery
[ ]
[ ] Car under dash
[ ]
[ ] Car reflector
[ ]
[ ] Closet bags, sporting equipment
[ ]

[ ] Closet, hang rod blocks
[ ]
[ ] Closet hang rod, hollow
[ ]
[ ] Closet shoes, inside
[ ]
[ ] Garage box, tool
[ ]
[ ] Garage containers, partially used
[ ]
[ ] Garage drain, water
[ ]
[ ] Garage rack, tool
[ ]
[ ] Garage battery charger
[ ]
[ ] Kitchen rolling pin, hollow
[ ]
[ ] Kitchen cupboards
[ ]
[ ] Kitchen canisters, dry food
[ ]
[ ] Kitchen container, food, animal
[ ]
[ ] Kitchen container, food, human
[ ]
[ ] Kitchen gourds, hollow
[ ]
[ ] Kitchen cans, aerosol, hollow
[ ]
[ ] Kitchen doors, pet
[ ]
[ ] Kitchen dishes, cups, glasses
[ ]
[ ] Kitchen jars, spice
[ ]
[ ] Kitchen ironing board
[ ]
[ ] Kitchen paper towel rack
[ ]
[ ] Kitchen cans, food
[ ]
[ ] Kitchen brooms, mops
[ ]
[ ] Kitchen racks, spice
[ ]
[ ] Kitchen canisters, condiments
[ ]
[ ] Living room headphones, stereo
[ ]

[ ] Living room fireplace screen
[ ]
[ ] Living room books, behind
[ ]
[ ] Living room VCR, inside, behind
[ ]
[ ] Living room ornaments, Xmas tree
[ ]
[ ] Living room doorbell chimes
[ ]
[ ] Living room cassette box, audio, video
[ ]
[ ] Living room plate, wall, cable TV
[ ]
[ ] Living room speakers, behind grill
[ ]
[ ] Living room stairs, posts
[ ]
[ ] Living room humidor, tobacco
[ ]
[ ] Living room cage, bird
[ ]
[ ] Living room aquarium, behind, under
[ ]
[ ] Living room stairs, banisters
[ ]
[ ] Living room posts on stairs, hollowed out
[ ]
[ ] Living room TV cable converter
[ ]
[ ] Living room vases
[ ]
[ ] Living room fireplace, flue opening
[ ]
[ ] Living room cassettes, audio, video
[ ]
[ ] Living room books, hollowed out
[ ]
[ ] Living room TV, remote control
[ ]
[ ] Living room, TV antenna, inside base
[ ]
[ ] Living room beverage cans, inside
[ ]
[ ] Living room pipe rack
[ ]
[ ] Living room fireplace, chimney
[ ]
[ ] Living room trophies and awards
[ ]

[ ] Office cabinet, filing
[ ]
[ ] Office desk under, hollow block
[ ]
[ ] Office computer, monitor, behind
[ ]
[ ] Office computer, keyboard, under
[ ]
[ ] Office computer, monitor, under
[ ]
[ ] Office fax machine, in, under
[ ]
[ ] Office printer, inside, under
[ ]
[ ] Office typewriter, inside
[ ]
[ ] Office computer, disk storage trays
[ ]
[ ] Office computer, keyboard, inside
[ ]
[ ] Office typewriter, behind
[ ]
[ ] Office pencil holder
[ ]
[ ] Office computer, case, inside
[ ]
[ ] Office Rolodex
[ ]
[ ] Office adding machine inside, under
[ ]
[ ] Office light fixtures, desk
[ ]
[ ] Office computer, modem, in, behind
[ ]
[ ] Perimeter pet houses
[ ]

[ ] Perimeter doorbell switch
[ ]
[ ] Perimeter phone, telephone pole
[ ]
[ ] Perimeter light fixtures
[ ]
[ ] Perimeter BBQ grill
[ ]
[ ] Perimeter doors, screen, hollow
[ ]
[ ] Perimeter furniture, patio
[ ]
[ ] Perimeter light, porch
[ ]
[ ] Perimeter box, connect, cable TV
[ ]
[ ] Perimeter phone connection box
[ ]
[ ] Perimeter BBQ tools, handles
[ ]
[ ] Perimeter lantern, Japanese
[ ]
[ ] Perimeter box, mail
[ ]
[ ] Perimeter window ledges
[ ]
[ ] Utility panel, electrical service
[ ]
[ ] Utility circuit breaker, unused
[ ]
[ ] Utility vacuum cleaner, inside
[ ]
[ ] Utility vacuum cleaner, handle
[ ]